LS Gen IV ENGINES 2005–Present

HOW TO BUILD *MAX PERFORMANCE*

Mike Mavrigian

CarTech®

CarTech®, Inc.
6118 Main Street
North Branch, MN 55056
Phone: 651-277-1200 or 800-551-4754
Fax: 651-277-1203
www.cartechbooks.com

© 2018 by Mike Mavrigian

All rights reserved. No part of this publication may be reproduced or utilized in any form or by any means, electronic or mechanical, including photocopying, recording, or by any information storage and retrieval system, without prior permission from the Publisher. All text, photographs, and artwork are the property of the Author unless otherwise noted or credited.

No portion of this book may be reproduced, transferred, stored, or otherwise used in any manner for purposes of training any artificial intelligence technology or system to generate text, illustrations, diagrams, charts, designs or other works or materials.

The information in this work is true and complete to the best of our knowledge. However, all information is presented without any guarantee on the part of the Author or Publisher, who also disclaim any liability incurred in connection with the use of the information and any implied warranties of merchantability or fitness for a particular purpose. Readers are responsible for taking suitable and appropriate safety measures when performing any of the operations or activities described in this work.

All trademarks, trade names, model names and numbers, and other product designations referred to herein are the property of their respective owners and are used solely for identification purposes. This work is a publication of CarTech, Inc. and has not been licensed, approved, sponsored, or endorsed by any other person or entity. The Publisher is not associated with any product, service, or vendor mentioned in this book and does not endorse the products or services of any vendor mentioned in this book.

Edit by Paul Johnson
Layout by Monica Seiberlich
ISBN 9781613253908
Item No. SA413

Library of Congress Cataloging-in-Publication Data

Names: Mavrigian, Mike, author.
Title: LS Gen IV engines 2005-present : how to build max performance / Mike
Mavrigian.
Description: Forest Lake, MN : CarTech Books, [2018]
Identifiers: LCCN 2018006751 | ISBN 9781613253908
Subjects: LCSH: General Motors automobiles--Motors--Performance--Handbooks, manuals, etc. | General Motors automobiles --Motors--Modification--Handbooks, manuals, etc. | General Motors automobiles--Motors--Customizing--Handbooks, manuals, etc. | LCGFT: Handbooks and manuals.
Classification: LCC TL215.G4 M34 2018 | DDC 629.25/04--dc23
LC record available at https://lccn.loc.gov/2018006751

Written, edited, and designed in the U.S.A.
Printed in China
10 9 8 7 6 5 4

CarTech books may be purchased at a discounted rate in bulk for resale, events, corporate gifts, or educational purposes. Special editions may also be created to specification.
For details, contact Special Sales at 6118 Main Street, North Branch, MN 55056 or by email at sales@cartechbooks.com.

DISTRIBUTION BY:

Europe
PGUK
63 Hatton Garden
London EC1N 8LE, England
Phone: 020 7061 1980 • Fax: 020 7242 3725
www.pguk.co.uk

Australia
Renniks Publications Ltd.
3/37-39 Green Street
Banksmeadow, NSW 2109, Australia
Phone: 2 9695 7055 • Fax: 2 9695 7355
www.renniks.com

Canada
Login Canada
300 Saulteaux Crescent
Winnipeg, MB, R3J 3T2 Canada
Phone: 800 665 1148 • Fax: 800 665 0103
www.lb.ca

CONTENTS

ACKNOWLEDGMENTS

Thanks to the following for the help in gathering the information provided in this book: Scott Gressman, Gressman Powersports; Bill McKnight, Mahle; Jason Harding, Katech Engines; Brian Carruth; Dick Bower, Erson; Trent Goodwin, Comp Cams; Jason Haines, Lingenfelter Performance; Justin Galvan, Lingenfelter Performance; Duane Boes, Callies Crankshafts; John Callies, Morel Lifters; Richard Maskin, Dart Heads; Jim Brewer, Dart Heads; Tim Torrecarion, AFR; Bill Mitchell, BMP; Russell Lehart, Mast Motorsports; Eric Blakely, Edelbrock; Phil Elliot, T&D Machine; Al Rebescher, Summit Racing; Dennis Ventrello, Jesel; Tony Lombardi, Ross Racing Engines; David Fussner, Wiseco Pistons; Sean Crawford, JE Pistons; Thor Schroeder, Moroso; Bill Tichenor, Holley; Dave Monyhan, Goodson Tools; C. J. Jones, Jones Racing Products; John Schwarz, AVIAID; and Leo Rayas, Barnes Systems.

Part I

LS Gen IV

In 2005, GM released the Gen IV engine family, which contains many upgrades over the Gen III. Increased displacement and several new features, such as better high-RPM flowing heads and updated camshaft sensing technology, produced a significant step forward in performance. The Gen IV LS engine lineup included iron-block 4.8L, 5.3L, 6.0L, and 6.2L V-8s while the cast-aluminum-block LS engines offered 5.3L, 5.7L, 6.0L, 6.2L, and 7.0L displacements. The stock LS Gen IV engines power the Camaro, Corvette, and other high-performance cars, SUVs, and trucks. In a certain sense, the LS is the perfect marriage of traditional cam-in-block engine architecture along with cutting-edge head design, modern electronics, and advanced cam technology.

While the Gen IV engines deliver exceptional stock-class performance, a myriad of high-performance parts and procedures can unlock the true potential of these engines. Part I provides detailed information and insightful instruction so you can build an engine to fit your specific output targets and budget. Guidance is provided for choosing a complementary performance package and determining the best machining procedures. In particular, this part covers engine blocks, crankshafts and connecting rods, cylinder heads, pistons, camshafts, lifters, rocker arms, oiling systems, and EFI and intake systems. In Part II, the author provides detailed and revealing instruction for completing two high-performance Gen IV engine builds.

LS Gen IV Engine Overview

The LS engine family began with what GM refers to as the Gen III engine platform, which included the LS1 and LS6. The Gen IV family began with the LS2 and includes LS7, LS3, LS9, and L92. Although the LS2 engine shared the cathedral intake port head design that carried over from the earlier LS engines, the LS2 falls under the Gen IV category basically because the camshaft timing sensor moved to the front of the block, whereas earlier blocks featured the cam sensor at the rear. This is the only reason the LS2 is referred to as a Gen IV engine.

All Gen IV LS engines feature a front-mounted cam position sensor (mounted to the front timing cover), with cam timing picked up by reluctor blocks on the camshaft gear.

LS2 Differences

The primary differences between the LS2 and other Gen IV engines involve increased displacement and cylinder head design. The LS2 offers 6.0L of displacement while the LS3 and LS9 feature 6.2L, with the LS7 providing 7.0L of displacement.

Of the various differences in cylinder head design, the most notable involve the intake ports. The LS2 heads feature the tall cathedral ports (similar to the intake ports found on LS1 and LS6), while the LS7, LS3, and LS9 heads feature a rectangular intake port design (often referred to as square ports, although they are actually rectangular).

This ready-to-install LS aftermarket crate engine featuring LS3 heads is available in horsepower ranging from 540 to 660. Thanks to LS cylinder head design and the appropriate cam profile, such horsepower numbers have become commonplace.

Performance Gains

The performance automotive aftermarket's prompt and enthusiastic support for the LS platform made horsepower gains possible through wide-ranging development and by offering stronger forged crankshafts, varying crankshaft strokes, stronger

This LS7 engine is being fitted to a 1969 Camaro. While creative turbo plumbing was required during fabrication and test fitting, engine and transmission installation was relatively straightforward thanks to readily available LS-swap mounts.

It's become rather commonplace for dual-turbo LS engines to produce well beyond 1,000 hp.

Aftermarket forged-aluminum pistons are offered for LS applications in a full range of diameters, compression heights, and dome volumes and configurations to suit any conceivable requirement.

The majority of LS factory engines feature powdered metal main caps. While suitable for street and some racing applications, a move to steel-billet main caps provides added bottom-end strength. Aftermarket blocks are commonly available with either ductile-iron or steel-billet caps.

Some factory LS engines, such as the LS7, are already fitted with steel-billet main caps. These OEM four-bolt main caps provide the strength to support more than 500 hp and 481 ft-lbs of torque.

forged connecting rods in lengths suitable for varying stroker combinations, and forged-aluminum pistons in a wide range of bore diameters, compression heights, and dome volumes. Roller camshaft profiles are available from all leading aftermarket cam makers to suit any desired build in terms of horsepower, torque requirements, and power bands. To provide much more durable threaded fasteners for the LS, stronger billet-steel main caps, cylinder head bolts or studs, main cap bolts or studs, and rod bolts are offered, all designed for torque-value installation, eliminating the need for torque-plus-angle tightening.

Based on the original GM LS cylinder head designs, sophisticated aftermarket cylinder head innovations have been developed that boost power well beyond that offered by the factory heads. Choices include Gen III and LS2 cathedral-port heads and Gen IV rectangular-port heads, with intake volumes ranging from about 225 cc to more than 255 cc, along with premium stainless-steel valves and improved flow designs for both straight and offset intake rocker applications and fully CNC-machined ports and chambers. The primary "secret" of LS power lies in the cylinder head designs. While factory heads such as the L92 style are excellent platforms, the aftermarket has really stepped up to provide power well beyond the OEM offerings.

LS cylinder heads are critical power makers due to their superior flow. Aftermarket head manufacturers have developed and refined the LS designs for even better flow by means of cathedral or rectangular intake ports and a wide range of valve sizes and combustion chamber volumes.

OEM rocker arms perform

Aftermarket forged cranks provide added strength and are available in a variety of strokes. Both standard and super-lightweight versions are offered, covering the needs of both street and extreme racing applications.

If you plan to reuse the factory rockers, you must upgrade to aftermarket fully caged bearing trunnions; factory uncaged bearings can dislodge.

Aftermarket full-roller rocker arms offer superior strength and reduced friction because roller bearings contact the valves instead of the OEM frictional contact design. Aftermarket performance roller rockers are available in both 1.72:1 and 1.8:1 ratios.

The aftermarket has stepped up to offer both iron and aluminum LS blocks that provide more latitude in overboring and stroking, in addition to enhancements that provide superior oiling, cooling, and, most important, strength.

The LS builder is no longer limited to the factory plastic lifter guide trays. Also available are tie-bar link lifters that provide superior precision by keeping lifter rollers in plane with the cam lobes. High-quality aftermarket lifters also offer superior strength for higher valve spring pressures and to accommodate higher engine RPM.

well, but aftermarket performance full-roller rockers enhance performance and durability with forged-aluminum bodies, hardened pushrod cups, heavy-duty caged trunnion bearings, and roller bearing valve tips. In comparison, the factory powdered metal rockers, which feature roller bearing trunnions, have frictional contact with valve stem tips. Aftermarket full-roller rockers are better able to withstand high engine speeds and higher cam lifts while offering reduced friction due to the roller bearing contact at the valves. Aftermarket rockers are also readily available in both a standard 1.72:1 ratio and a 1.8:1 ratio.

Retrofitting the LS

While factory lifters are individual and are guided within plastic guide trays, aftermarket choices now include lifters that retrofit to these plastic guides and link-bar-paired roller lifters that eliminate the plastic guides. In basic terms, the aftermarket fulfills the requirements of performance and racing applications by addressing the various weak points of the factory designs both to boost power and to provide a much higher level of durability for high-stress application environments.

A full range of additional components is readily available to increase power and durability, including viscous-dampened crankshaft balancers, improved-flow mechanical and electric water pumps, and enhanced-design timing covers that provide additional timing chain clearance and allow cam timing changes without the need to remove the crank pulley. Improved oil pan designs are now available that provide added capacity and reduce windage concerns. Also offered are more-robust and powerful ignition coils, taller valve covers that provide clearance for aftermarket rockers, adapters that permit installation of Gen I small-block Chevy valve covers, a variety of intake manifolds for both injected and carbureted applications, and easy conversions to allow the use of carburetors on LS engines. Forced induction systems are also readily available, including both supercharged and turbocharged applications.

Unlike the early days when the LS platform was considered the "new and unique" performance engine, thanks to the performance aftermarket we now have total coverage in performance upgrades; that is,

Electric water pumps are popular among many builders and are now available for LS applications. Depending on your belt routing requirements, these pumps are offered with or without an idler pulley.

Aftermarket high-performance ignition coils offer superior output, reliability, and appearance over factory coils.

Factory connecting rods on most LS engines are powdered metal construction. The factory rods are suitable for power up to around 450 hp or so, but it is strongly recommended to swap to forged-steel aftermarket rods, which are available in various center-to-center lengths to accommodate standard or stroker applications.

for *every* single component involved in a build. Regardless of your goal, whether that involves a mild street upgrade or full-blown competition in any form of racing, including drags, road racing, oval track, etc., if you've got the money, the aftermarket can feed your addiction.

Although factory blocks can handle minor to major power enhancements, aftermarket blocks in both cast iron and aluminum offered by manufacturers such as Dart, Racing Head Service (RHS), and World Products provide notable improvements and versatility. Aftermarket blocks allow builders to obtain larger cylinder bore diameters and accept longer crank strokes, along with vastly improved strength and durability, giving engine builders much greater latitude in achieving more power, torque, and reliability. Thanks to support from the performance aftermarket, achieving naturally aspirated 600-plus hp is extremely easy, with forced-induction builds reaching 1,000 hp and beyond.

For those who wish to run their LS engines carbureted, a variety of intake manifolds are available in dual-plane, single-plane, low- and tall-profile, and single- or multiple-carb designs. These manifolds are also offered without injector bungs or with injector bungs for those who wish to run a central-mounted throttle body with per-cylinder injectors.

Example of a custom-built LSX LS7 street/competition engine featuring a FAST intake manifold. (Photo Courtesy Livernois Engines)

Upgrading critical fasteners in cylinder head, main cap, and rod bolt applications is essential for any high-horsepower build. Not only do quality aftermarket fasteners provide superior tensile strength, they are specified to use a torque value only, an easier installation than the factory torque-plus-angle tightening method. Unlike factory fasteners that are torque-yield one time use, quality performance aftermarket fasteners may be reused, obviously based on condition.

Instead of using the factory crankshaft pulley, aftermarket balancers are available with or without grooved pulleys. The viscous-dampened balancers provide additional reduction of harmonics. These are available for either unkeyed factory-type crank snouts or keyed for aftermarket cranks.

Because of the increased popularity of the LS platform, more and more restoration builders choose LS transplants for older muscle car applications, spawning a range of engine swap kits. Components developed to ease installation of the LS engine into older vehicles include motor mounts, frame adapters, plug-and-play controller systems for factory-type injection applications, exhaust systems that allow LS engine adaptations into a variety of popular muscle car and vintage vehicles. Essentially, all the types of power-adding and increased-durability components developed over the decades for Gen I engines are now available for the LS platform, making it the most popular standard for Chevy performance applications today.

Miscellaneous Details and Differences

- All LS-family engine blocks feature a 4.40-inch cylinder bore spacing (bore center to bore center).
- LS3 and LS9 engine blocks are similar. However, head bolt holes differ: LS2, LS7, and LS3 feature 11-mm head bolt threads, while the LS9 features a stronger 12-mm diameter.
- LS3 and LS9 have a 4.06-inch bore and 3.62-inch stroke for 6.2 liters of displacement.
- LS7 cylinder walls are considered too thin for forced-induction applications.
- LS3 and LS9 heads are similar, but LS9 heads are cast from a special A356-T alloy to better accommodate high heat conditions, which also helps to eliminate casting porosity concerns.
- LS9 heads have intake ports with "swirl wings" for better combustion efficiency.
- LS3 and LS9 coils are identical, but LS9 ignition coils are mounted directly to the valve covers to save space.
- LS2, LS7, and LS3 factory-built engines utilize two-layer multi-layer steel (MLS) head gaskets, while the LS9 was fitted with four-layer MLS head gaskets. A variety of MLS gasket configurations are now available for all LS platforms, including gaskets with larger bore sizes to accommodate modified larger cylinder bores, as well as variations of MLS cylinder head gaskets that allow tailoring to desired compression ratios.
- LS9 engines feature forged pistons at a 9.1:1 compression ratio as opposed to hypereutectic pistons. Titanium connecting rods are featured in the LS7. The LS9 also features a forged-steel crankshaft and piston oil squirters that are located just below piston bottom dead center (BDC) for added lubrication delivery to the piston skirts.
- LS7 aluminum blocks feature press-fit iron cylinder sleeves as opposed to integrally cast-in liners, 4.125 bore/4.000 stroke, a forged 4140 crank, titanium connecting rods, 11:1 hypereutectic pistons, doweled billet-steel main caps as opposed to powdered metal main caps, .600 cam lift with 211/230 duration at .050 inch, dry sump oiling for improved ground clearance and more immediate and consistent oil delivery in aggressive handling and acceleration conditions, 12-degree CNC-ported heads, and titanium intake valves.
- The majority of factory LS engines feature a crankshaft stroke of 3.622 inches.
- Truck engines in the LS family are referred to as Vortec engines. Initially, engines for truck applications featured only cast-iron blocks, but this was later transitioned to either iron or aluminum blocks.
- Gen IV engines, depending on vehicle applications, featured a provision for GM's variable valve timing (VVT) or displacement on demand (DOD) systems, wherein some cylinders were deactivated during low-demand use to increase fuel economy. These systems are often disabled or deleted by enthusiasts for performance use.

Engine Blocks

Engine blocks in the LS family share similar characteristics, with deviations primarily based on cylinder bore size. One of the primary differences between Gen III and Gen IV blocks is the relocation of the camshaft position sensor, which was moved from the Gen III location at the rear top of the block to the timing cover on Gen IV engines. All factory production blocks, whether cast iron or aluminum, feature six-bolt main caps, with four primary vertical bolts and one 8-mm side bolt at each side of each main cap. All LS production blocks feature powdered metal main caps except LS7 and LS9 engines, which feature steel-billet main caps for added strength. In comparison to earlier blocks, LS3 and LS9 blocks feature additional reinforcement in the main web areas. LS9 blocks were designed for supercharging forced induction, so they feature larger bulkhead windows for improved bay-to-bay breathing; larger-diameter 12-mm head bolts; and piston oil squirters in the cylinders.

Block Specs

Type	Displacement	Material	Weight	Bore Diameter	Stroke
LS1/LS6	5.7L/346 ci	aluminum	110 pounds	3.898 inches	3.622 inches
LQ9	6.0L/364 ci	iron	170 pounds	4.000 inches	3.622 inches
LS2	6.0L/364 ci	aluminum	110 pounds	4.000 inches	3.622 inches
LS3/L92	6.2L/376 ci	aluminum	110 pounds	4.065 inches	3.622 inches
LS7	7.0L/427 ci	aluminum	110 pounds	4.125 inches	4.000 inches
LS9	6.2L/376 ci	aluminum	110 pounds	4.060 inches	3.622 inches
LQ4	6.0L/364 ci	iron	170 pounds	4.000 inches	3.622 inches
LS4	5.3L/325 ci	aluminum	110 pounds	3.780 inches	3.622 inches
LY6	6.0L/364 ci	iron	170 pounds	4.000 inches	3.622 inches
L76	6.0L/364 ci	aluminum	110 pounds	4.000 inches	3.622 inches

Block Part Numbers	
LS2	12568950
LS3	12584727
LS7	19213580
LQ9	12572808

Stock Blocks

Factory and aftermarket blocks are substantially different from one another. Factory blocks are mass-produced with wide tolerances permitted as "acceptable" for common street applications. For example, while the factory specification for LS block deck height is 9.240 inches, OEM blocks rarely meet this spec. It's rather common for a block to feature greater and/or lower deck height on any particular block, with deck height varying from low to high along either bank. This means that the decks may not be parallel to the crankshaft centerline. While this may not present a problem for the average street engine, if your goal is to obtain maximum power, decks must be checked and likely corrected to achieve the same cylinder volume for all cylinders. Due to shifts in block geometry that result from the casting process, some cylinder walls may be thinner than others. The list goes on, but you get the point. If you wish to achieve maximum results for a power build, a factory block will likely require several corrective machining operations to "accurize" the block.

Unlike other LS blocks that feature powdered metal main caps, the LS7 comes from General Motors with steel-billet main caps.

The LS7 features pressed-in cast-iron cylinder liners that are the biggest cylinders in the LS lineup. The race program provided data that was used to develop a light, rigid block. The deep-skirt configuration provides exceptional strength. The bulkheads hold six-bolt, cross-bolted main bearing caps that mitigate crank flex. This is the front view of an LS7 block.

The LS2 features a 4.000-inch bore and a 3.622-inch stroke. In contrast to earlier Gen III blocks, the LS2 cylinder head bolt holes are blind and not open to water. Note the four-lug cam timing reluctor for the front-mounted cam position sensor on this cutaway of the LS2 6.0L aluminum block, deviating from the Gen III rear-mounted cam sensor.

The rear view of an LS2 block cutaway shows the bay-to-bay breathing openings at the bottom of the cylinders, typical of all LS blocks. The LS2 block is very similar to earlier blocks, with changes made to the cam position sensor location and blind head bolt holes.

In contrast, high-performance aftermarket blocks feature design enhancements such as thicker decks, stronger main webs, improved cylinder cooling, and improved oiling circuits, in addition to more precise CNC machining. Granted, a new bare block will need to be final-machined for the desired block deck height, lifter bore and cylinder bore diameters, and other dimensions. This final machining is by design: The manufacturer provides extra material that allows you to custom-fit your specific components. In addition, aftermarket block makers provide much greater attention to detail in terms of raw machining, wherein crankshaft centerline, camshaft centerline, lifter bore spacing, and cylinder bore centerlines are already spot-on. While correcting an OEM block may force you to compromise in terms of certain dimensions, with a quality aftermarket block you can obtain exactly what you want.

Even though factory blocks have been proven to handle drastic horsepower increases, such as builds configured with single or twin turbochargers that produce 1,000 hp and beyond, longevity and durability are key issues. These blocks were not designed to withstand the level of high combustion pressures associated with such power levels. In my opinion, and in the opinion of many other engine builders, if you plan to increase power this dramatically, upgrading to a stronger aftermarket block greatly reduces the risk of catastrophic failures that could result from excessive cylinder bore distortion and over-stressed main webs. If a build is planned to produce more than 700 hp, a stronger and beefier aftermarket performance block provides a much more reliable and stable platform for the build. Aftermarket performance and racing blocks feature thicker decks, stronger main webs, enhanced priority main oiling systems, thicker cylinder walls, vastly improved bay-to-bay crankcase breathing, and stronger-grade materials in both alloy and iron configurations.

Modifying the Stock Aluminum Block

Modifying a stock block can involve both corrective and enhancement processes. It's very common for stock LS blocks to have uneven and out-of-specification decks. While the spec deck height is 9.240 inches, you may find blocks that have slightly taller or shorter decks. In addition, LS factory blocks tend to be out of square, with the front or rear of the decks being shorter or taller than the opposite decks. The decks can be resurfaced to make them the same height and parallel to the main bore centerline, using the shortest area as the index. Making the decks parallel to the main bore centerline helps equalize the combustion area between the piston at TDC and the cylinder head combustion chambers, as well as equalizing pushrod length requirements from cylinder to cylinder.

Performance aftermarket blocks usually provide a bit of extra deck height, allowing you to achieve the desired deck height. Finishing to the desired deck height will allow you to make the decks parallel to the main centerline.

While GM LS blocks feature main caps secured with two primary bolts and two side bolts, some aftermarket blocks eliminate the side pinch bolts and feature a four-bolt main cap design.

If you intend to increase displacement by moving to larger cylinder bores, be aware that factory aluminum blocks have bore liners that are installed during the casting process. The liners are relatively thin, allowing an overbore of only about .005 inch to a maximum of about .010 inch. If you intend to go .010 inch oversize, a sonic wall-thickness gauge should be used to measure wall thickness before any oversizing is performed. Factory iron blocks can be oversized more, again assuming that you'll have at least about .200-inch wall thickness once the bore has been machined. Factory iron blocks can routinely be larger than the specs given by .030 inch, with some capable of handling as much as a .060 inch oversize, again, only if cylinder wall thickness is not compromised. Not all factory blocks are identical due to core shift during the casting process, so each cylinder should be first checked for wall thickness.

Aftermarket blocks tend to provide thicker cylinder walls, potentially allowing larger oversizing, but always check with the block manufacturer for the bore diameter limitations. Always refer to the piston skirt diameter of the pistons that will be installed to determine the required piston-to-wall clearance. Piston-to-wall clearance can vary depending on the piston material and the intended use (street, street/strip, race, forced induction, etc.). Never finish cylinder bores unless you have the intended pistons in hand so that pistons can be measured for skirt diameter.

Lifter bores should also be checked both for diameter and for angle. Depending on the lifters being used, desired clearance can vary, with advised clearances of some aftermarket lifters in the .0015-inch area. Always measure lifter diameter and verify that lifter bores are sized appropriately for the lifters that will be installed. Although not extremely common, due to potential core shift in factory blocks, lifter bore angles may be not perfectly perpendicular to the camshaft. Using specialty aftermarket accurizing fixtures or with the use of CNC machining, lifter bores can easily be corrected if needed. If the process of correcting results in oversizing, bronze bushing can then be installed and machined to size.

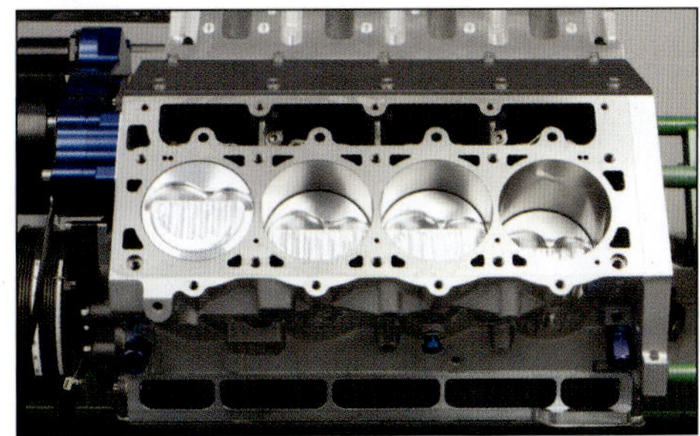

This view of a six-bolt deck clearly shows the extra head bolt holes above and below each cylinder, inline with the cylinder centerline.

Aside from these modifications, clearance checking when using a longer-stroke crankshaft is always required, with clearances measured between crank counterweights and the block pan rails, connecting rod big end to pan rails and cylinder bottoms, and rod big ends to camshaft lobes. Naturally, this is done during initial test assembly before any machine work is done.

Disabling Performance-Robbing Displacement on Demand

For high-performance LS Gen IV engines, the displacement on demand (DOD), also called active fuel management (AFM), must be disabled because it will restrict performance. Many of the LS Gen IVs use this system, except for the LS7 block. It disables four cylinders during certain driving conditions, such as easy cruising with low engine load. The only purpose of DOD/AFM is to increase fuel economy, a factor that isn't high on the priority list for most performance-minded owners.

When DOD is signaled, cylinders 1-4-6-7 are shut down by effectively disabling their valves and by cutting off the spark to their ignition coils. This happens when solenoids on the underside of the valley cover allow high-pressure oil to be delivered to a groove in the special two-part DOD lifter that causes a pin to collapse a spring-loaded pin in the lifter. The two parts of the DOD lifter are then free to move relative to one another. The cam lobe keeps pushing on the lifter follower, but the inner part of the lifter pushes against a coil spring at the top of the lifter to prevent the force from being transferred to the pushrod, so that the rocker arm does not push on the valve. When the DOD system turns on, the V-8 engine effectively becomes a 4-cylinder engine, saving fuel.

When more power is required, the lifters are activated. The operation is controlled by the engine control unit (ECU) and four solenoids located in the lifter valley. The solenoids provide a pressurized oil signal to the roller lifters. The system is tuned to a specific camshaft profile, with different lobe profiles between AFM and non-AFM cylinders and different valve lash requirements. DOD/AFM lifters were utilized in various 2006–2015 Gen IV LS engines. Examples include 5.3L engines in various vehicle applications, 6.2L L94 engines in Cadillac Escalades, 6.0L L76 engines in 2007–2009 Chevy Avalanches, 6.2L L99 engines in 2012–2015 Camaros, and 6.0L L77 engines in Chevy Caprice models.

The oiling towers of this LS7 block are blank and not drilled open. Since some engine applications feature DOD/AFM and some don't, a common casting was made, with towers drilled for oil passages when factory production called for the DOD/AFM feature.

If a more aggressive cam is installed, the DOD/AFM system must be disabled, requiring not only the cam change but also installation of different lifters and lifter guides. In addition, the ECU must be recalibrated. Any competent GM dealer's service department can

The oil passages in the DOD/AFM towers are open on this block, and at this stage, it's ready for tapping and plugging.

This factory LS7 block has the four-bolt-per-cylinder-head bolt layout, the siamesed cylinder bores, and the main cap side bolt holes. The siamesed bore provides additional strength to the block and the four-bolt head provides extra clamping force for high-horsepower output.

reprogram (reflash) the ECU to keep the computer from detecting the on-demand lifter deactivation. If this reflash isn't done, the driver will see a check engine light illuminated on the instrument cluster.

Note that the LS7 block features the same oil pedestals in the lifter valley, but they are not drilled open; there's no need to make modifications because this block is not equipped for DOD lifter deactivation.

Many builders elect to eliminate DOD/AFM because it is simply not needed when the goal is to gain full advantage of the engine's performance at all times. Eliminating DOD/AFM requires plugging all oil delivery ports in the standoffs located in the block's upper valley. If these ports are left open, internal oil leaks will result in low oil pressure. The oil ports in the valley can be plugged by drilling and tapping each port and installing 1/8-inch NPT plugs, but this should be done only on a bare block that will be properly washed and rinsed after machining operations.

The Lingenfelter rivet tool allows easy plugging of DOD/AFM oil ports in the valley stands. (Photo Courtesy Lingenfelter Performance)

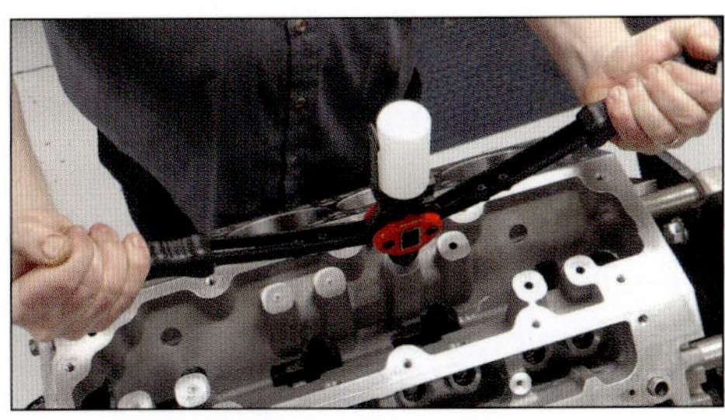

After installing a rivet to the tool, insert the rivet into the oil port with the tool arms spread apart. (Photo Courtesy Lingenfelter Performance)

With the rivet tool held against the stand, begin to squeeze the tool arms to begin rivet expansion. (Photo Courtesy Lingenfelter Performance)

Continue squeezing the tool arms until the rivet fully expands and the mandrel snaps free. (Photo Courtesy Lingenfelter Performance)

No additional sealant is required. The aluminum plugs seal the ports and will not loosen. If you wish to eliminate the DOD/AFM system, these holes must be plugged to avoid a reduction of oil pressure. If an aggressive camshaft is being installed, the DOD/AFM system must be disabled. (Photo Courtesy Lingenfelter Performance)

Rivet Plugs

Lingenfelter Performance offers a slick alternative that can be handled even on an assembled block. This essentially involves installing an aluminum rivet plug in each port with the use of a manual rivet tool. This requires no machining, so the concern about leaving metal particles and shavings inside the oil passages is eliminated.

To install a rivet, put the rivet into the rivet tool. With the tool's arms spread apart, insert the rivet into the oil port. While holding the tool against the port surface, squeeze the tool's arms together to expand the rivet. Once the rivet fully expands, the rivet mandrel will snap off and remain in the tool. The aluminum rivet plug will seal the port with no additional sealing required. Lingenfelter's installation tool is available as PN L950105305, which includes a set of rivets. Additional eight-rivet sets are also available as PN L960225305. This is a slick, easy, and no-mess method of sealing the oil ports on either a bare or assembled block.

DOD Delete Kits

Eliminating the DOD system isn't complicated. This involves closing off the oil passages in the DOD towers, replacing the four cylinders' intake and exhaust lifters with "regular" LS lifters, replacing the DOD-location plastic lifter guides with standard non-DOD lifter guides, replacing the camshaft with a non-DOD cam, and eliminating the solenoids by replacing the lifter valley cover. You can source all the items individually, or you can buy a DOD-elimination kit from General Motors or various aftermarket sources.

A GM Performance Parts DOD delete kit is available under GM PN 12570471. This kit includes a three-bolt 4X cam sprocket, ARP cam sprocket bolts, LS2 chain damper, LS7 lifter set, LS3/L92 head gaskets, OEM head bolts, exhaust manifold gaskets, LS2 valley plate, a set of four standard plastic lifter guides, PCV dirty air hose, PCV valve cover plug, and a new GM balancer bolt. Installing the kit also requires the use of an aftermarket camshaft that was not designed for use with DOD/AFM along with ECM reprogramming.

One example is Summit Racing's PN CMB-09-0026, which includes a set of lifters, lifter guides, valley cover, and head bolts. Another DOD/AFM delete kit is Tick Performance's PN 5065TP. This includes an LS3 valley cover, LS2/LS3/LS7/LS9/L92 lifter trays, a full set of LS7 lifters, head gaskets, head bolts, LS2/LS3/LS7/L92 timing chain damper, LS2 PCV hose, PCV cap, timing cover gasket, water pump gaskets, timing cover seal, and a GM crank balancer bolt.

While most factory production LS blocks feature powdered metal main caps (except the LS7, which features steel-billet caps) that use 8-mm side "pinch" bolts, some aftermarket blocks utilize steel-billet four-bolt main caps with outer splayed bolts, such as this example of a Dart LS Next block.

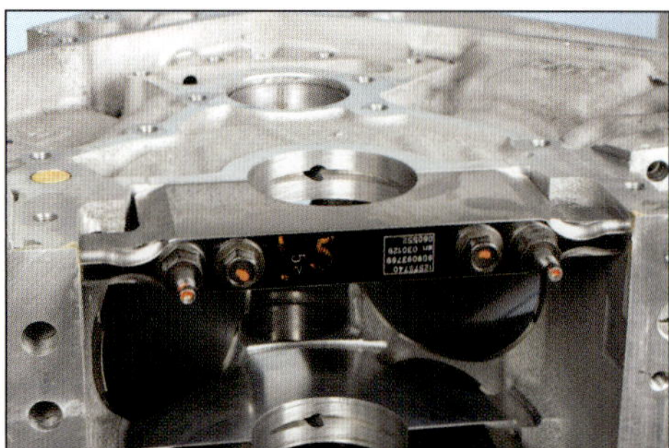

You can see the machined ends on the LS7 billet-steel main caps that accept the 8-mm side bolts. Doweled-in forged-steel main bearing caps provide excellent support for the crankshaft.

Main Caps

All GM factory LS blocks feature a six-bolt main cap design, with four primary 10-mm x 2.0 vertical bolts and two 8-mm x 1.25 "pinch" or "cross" bolts that pass through the lower block sides into the caps. All factory LS blocks also feature powdered-metal main caps except the LS7 and LS9 blocks, which use forged-steel main caps. Powdered metal caps have proven to be satisfactory for builds up to the range of about 500 hp. Beyond that horsepower range in a naturally aspirated engine, and especially if higher cylinder pressures are planned due to nitrous injection and/or forced induction, aftermarket forged-steel main caps along with stronger main bolts or studs, such as those offered by ARP and other sources, are highly recommended.

Aftermarket Blocks

If you want to up the ante for increased block strength and/or more displacement, several high-performance blocks are available in the aftermarket for radical street or all-out racing applications. Available in cast iron, cast aluminum, and even billet aluminum, these blocks are designed as an improvement of the factory LS design to accommodate higher-horsepower applications and better withstand the abuse of increased cylinder pressure at higher engine RPM, increased compression, and/or forced induction. Several designs also provide enhanced priority main oiling circuits and improved cooling jacket designs, along with available beefy billet-steel main caps. These aftermarket blocks offer increased durability for higher-demand applications.

Dart LS Next Block

Dart offers an array of LS-based blocks designed to provide superior power, reliability, and durability.

An aftermarket LS block alternative is Dart's LS Next block, with thicker decks and elimination of the Y-block extended pan rails for crank windage improvements. Full-skirted blocks are also available.

Several performance aftermarket blocks are available, allowing a stronger build for higher cylinder pressure and extended RPM applications. Shown here is Dart's LS Next block, which features thicker decks, priority main oiling, thicker cylinder walls, and a more-dense high-nickel casting. Similar to design features found in GM LSX race blocks, the decks offer six-bolt-per-cylinder-head fastener locations.

The inboard "extra" head fastener locations on the Dart block are smooth-bored, requiring studs that secure to the head deck. The studs pass through the open holes and are secured with shouldered washers and nuts from the valley side.

Some aftermarket LS blocks feature the six-bolt head-bolt design, with extra upper and lower head bolt holes added at each cylinder location, as shown on this Dart LS Next block. Head studs provide added strength.

Dart's LS Next block also features four-bolt steel-billet registered main caps, eliminating the need for main cap side bolts.

In addition to Dart's offerings of billet-aluminum and cast LS blocks, the company has also introduced its LS Next block, a further evolution of the LS format.

Dart's latest block is vastly improved compared to the factory block. It is offered in iron or aluminum; the LS Next block features radical enhancements that include elimination of the factory Y-block skirt for reduced windage and allowing for thicker, stronger full-main web architecture. In addition, the design incorporates extended cylinders that are .375 inch longer at the bottom for better piston skirt support at BDC; low-restriction priority main oiling system; and four-bolt main caps with 7/16-inch bolts. Deck heights are offered in both stock 9.240-inch and optional 9.450-inch sizes to allow displacement increases. Thicker siamesed bores provide for increased oversizing potential for added displacement. Thicker 5/8-inch decks have been added for rigidity. A larger water jacket is installed at the number-1 cylinder, and in the valley is a provision for oil restrictors.

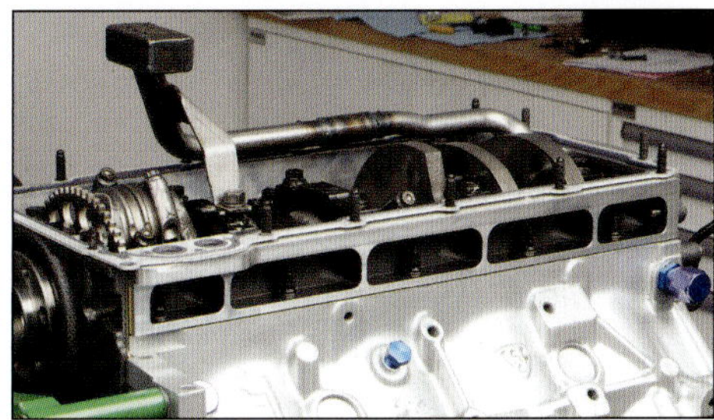

Because the Dart LS Next block's pan rails have been shortened by 2 inches to improve windage, accommodation is required to mount the oil pan. Choices include a custom pan from Canton or Stefs (these are designed with taller sides) or a pair of spacer rails that mount to the block, effectively regaining the needed mounting surface for the oil pan.

The LS Next block is designed to easily handle 1,500-plus-hp levels when the engine is pushed hard, as in racing conditions. I've seen reports

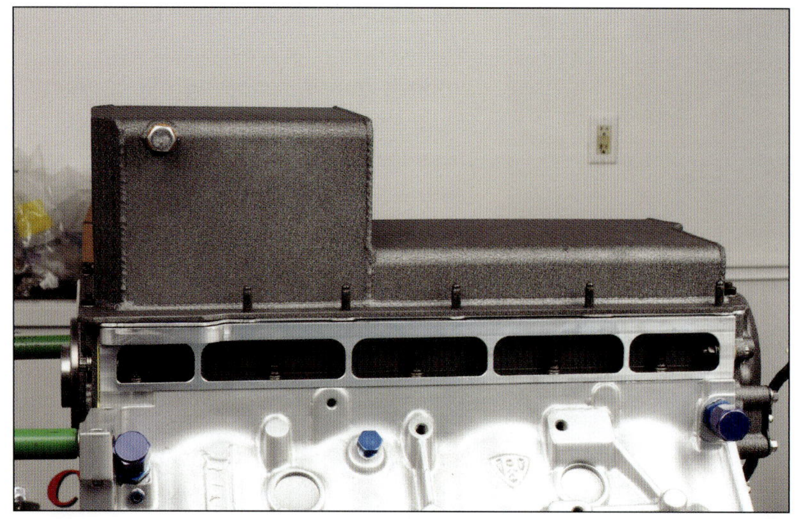

With the Moroso 2-inch rail spacers in place, an LS-style oil pan may be mounted. Here a Moroso pan is installed.

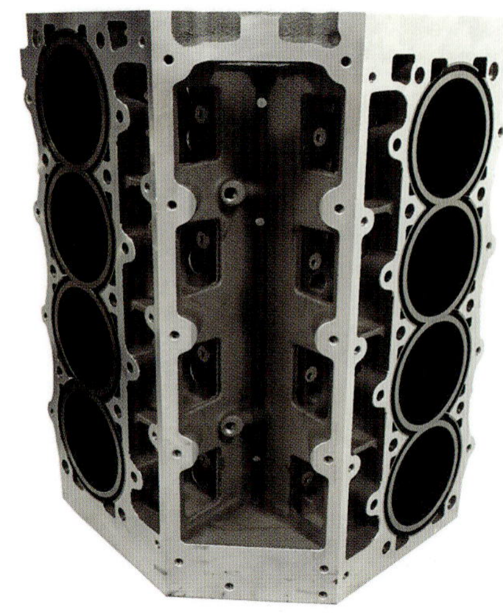

Dart's LS Next MID block features a modular integrated deck design that offers increased cylinder integrity and strength. According to Dart, the distortion is removed from the deck and the sleeves are anchored in compression in the lower block area for enhanced strength. (Photo Courtesy Dart Machinery)

of this block in applications pushing past 2,500 hp. These blocks are intended for extreme conditions in motorsports applications. Although I have not built an engine approaching the 2,000-hp level, I certainly view this type of block to be most suitable for extreme power levels, far beyond what a factory stock block could handle. Keep in mind that even the highest-grade block can fail because of problems with other component issues and/or incorrect assembly techniques, such as less-than-robust connecting rods, inadequate valve-to-piston clearance, improperly tightened rod bolts, inadequate bearing clearances, poor oil delivery, etc. But if you want a strong basis for a high-performance build, aftermarket blocks such as the LS Next offer vastly improved strength and durability. Blocks such as these provide a sturdy foundation for extreme power builds. Keep in mind that an engine block unto itself does not make more power; the design and strength incorporated into the block simply allow you to build more power and to better withstand extreme abuse.

Note: A special oil pan is required because of the elimination of the Y-block skirts. Both Canton and Stefs currently offer an appropriate pan, and Moroso carries a dedicated oil pan kit. The Canton and Stefs pans feature 2-inch-taller rails built into the pan design, while the Moroso kit uses 2-inch-thick spacers. This is required to compensate for the 2 inches that were removed from the block's pan rails.

Dart LS Billet Block

One of the many block designs offered by Dart is its billet LS block. The block is machined from a blank chunk of high-grade, aerospace-quality aluminum alloy, offering "virtually unlimited" choices of specifications, including bore centerline, bore diameter, deck height, lifter bore sizes, and cam tunnel height placement. Basically, it's custom machined to your spec

Dart's billet LS block is entirely machined from a chunk of high-grade aluminum alloy to finished state on CNC, offering an increase in strength as well as weight reduction. The OEM cast-aluminum bare block weight is about 110 pounds for LS1/LS6/LS3/L92. The weight of Dart's billet block is about 125 pounds, depending on deck height and bore size. The slight increase in weight compared to the OEM casting is inconsequential, given the billet block's vastly increased strength. As a reference, the OEM cast-iron LQ9 6.0L block weighs in at about 170 pounds. Shown here is a billet block ready for final cylinder honing to accommodate the builder's specific piston diameter. (Photo Courtesy Dart Machinery)

The recently introduced World Products Motown II LS iron block incorporates aftermarket-influenced and enhanced small-block Chevy architecture. It features a 9.240-inch LS deck height with a .134-inch raised cam bore, combining the strength of a performance-upgraded tried and true small-block Chevy with the vastly superior breathing of LS cylinder heads.

The most notable design features of the Motown II LS block are the LS cylinder head bolt locations, deck height, and cooling passages.

block, offered with either steel or aluminum main caps.

Dart LS MID Block

Dart's LS Next MID block features a modular integrated deck as well as a wet cylinder design to virtually eliminate cylinder distortion from block flexing and harmonics. Cylinder sleeves are seated and anchored in compression in the lower block area where the maximum aluminum mass is concentrated, offering increased cylinder integrity and strength compared to dry sleeves. Compatible with all currently available LS head designs, the Next MID is available in 9.240- and 9.750-inch decks heights and 4.125- to 4.220-inch bore sizes. It requires the use of MLS head gaskets.

World Products Motown II LS Block

This block represents a distinct departure from a typical LS design because the block is essentially a

The lifter bore locations are specific to the LS layout. Bushed lifter bores are standard and must be final-machined to accommodate the lifters of choice. Blocks can be ordered to accommodate either .847- or .904-inch lifters.

The mandatory aluminum valley cover base not only provides a cover for the lifter valley but the angled sides also extend the block's deck surface to complete the footprint for the LS cylinder heads and feature 5/16-18 threaded holes to accept the heads' inboard pinch bolts. The base also features a distributor mounting flange to accommodate conventional small-block Chevy distributor mounting. Since the valley cover base completes the deck surface area for the heads, this base must be mounted to the block during final deck surfacing.

Gen I small-block Chevy design. However, it accepts LS cylinder heads, crankshaft, rods, oil pan, and distributor, but the distributor can be eliminated if you prefer to use a controller and LS coil packs. It also accepts a water pump, crank balancer, oil pump, and other components. The block's deck height is 9.420 inches with an LS head bolt pattern to allow the use of LS-style heads. The LS-style components include cylinder heads, LS-style pistons, rockers, lifters, and intake manifold.

The block requires the use of a special camshaft that is essentially a small-block Chevy cam (with distributor drive gear) that features LS cam lobe spacing. The camshaft, which must be custom ordered to meet lift, duration, and LSA requirements, is currently offered by Comp Cams and Erson. Don't be alarmed at the "custom order" aspect: normal delivery time is only about a week or so. This block design is intended to offer the best of both worlds, combining the easy access and flexibility to accept more affordable tried and true Chevy small-block components with the vastly improved performance of the LS cylinder head designs. A special aluminum valley cover base is available with or without a distributor mounting hole. A separate flat aluminum valley cover plate seals the lifter valley.

One "unusual" aspect of the block is the use of two external coolant hoses that run from each side of the block to a high-mounted water housing. Each side of the block features a threaded water jacket hole that accepts a 1¼-inch NPT male thread fitting that accepts a –12 AN 90-degree hose end. The –12 AN hoses then connect to the coolant fill reservoir that is mounted to the front of the block. The housing features a hose neck for upper radiator connection.

In fact, I recently built a 427-ci engine using one of these blocks. It featured a 4.125-inch bore and 4.000-inch stroke. Components included a Scat 4.000-inch stroker crank, Scat 6.125-inch rods, JE pistons at 1.115-inch compression height, .624-inch-lift Erson hydraulic roller cam, Trick Flow LS CNC heads, Holley single-plane intake manifold and 850-cfm carb, MSD distributor, Fluidampr balancer, Melling oil pump, Moroso pickup and pan, Cloyes timing set, Comp aluminum roller rockers, Trend 7.500-inch pushrods, and ARP main and head stud kits. On the dyno, it easily pulled 641 hp and 555 ft-lbs of torque.

One of the cool features of the block is the motor mount bolt hole locations: both small-block Chevy and LS patterns are built in, allowing an easy swap into any vehicle that was originally intended for a Gen I Chevy or an LS engine.

As is the case with most aftermarket performance blocks, this block is already notched for rod clearance, in this example for a 4.000-inch stroker crank. Additional notch clearancing may be needed for longer strokes.

When ordering a block, options include a choice of either modular or steel-billet main caps. The steel-billet 3, 4, and 5 caps feature splayed outer bolt locations.

RHS Race Block

RHS's LS Race Block, cast from A357T aluminum alloy and CNC-machined, is intended for all-out race applications. Press-in spun-cast extra-long cylinder bore liners can be finished to 4.125- to 4.165-inch bore diameters. Available in standard LS 9.240- or tall 9.750-inch deck height, the block is designed to accept up to a 4.600-inch stroke and can accommodate up to a 60-mm roller bearing cam. The cam centerline is raised .388 inch to allow for longer stroke. Based on LS7 design, the decks feature a six-bolt-per-cylinder layout with a full water jacket around each cylinder. The block will accept either LS or early generation motor mounts (bellhousing mounts will also accept early or late bellhousings).

As a convenience feature, both small-block Chevy and LS motor mount bolt holes are provided, making it easy to install this block in an early vehicle that originally had a small-block Chevy engine or a later vehicle that was equipped with an LS engine; no custom motor mounts are needed.

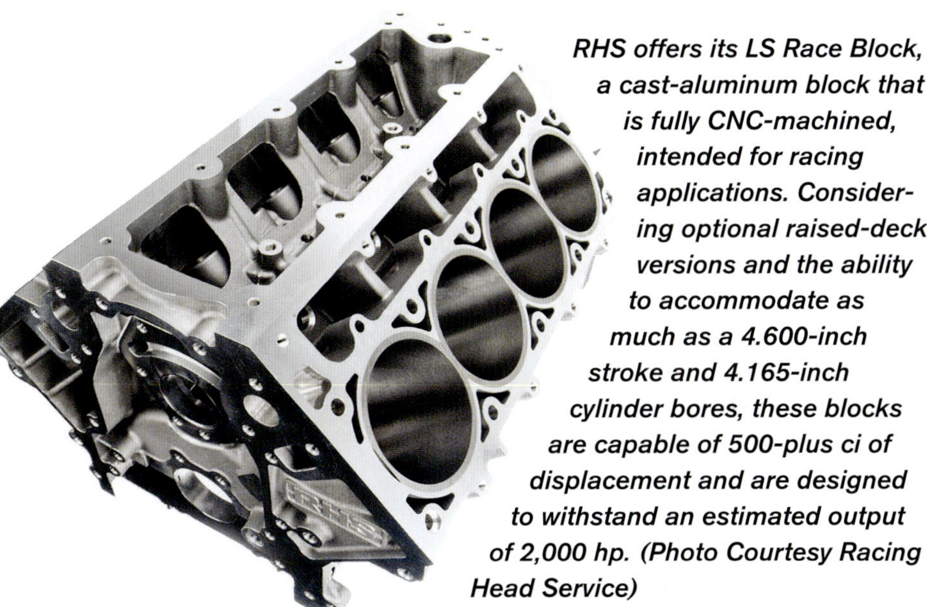

RHS offers its LS Race Block, a cast-aluminum block that is fully CNC-machined, intended for racing applications. Considering optional raised-deck versions and the ability to accommodate as much as a 4.600-inch stroke and 4.165-inch cylinder bores, these blocks are capable of 500-plus ci of displacement and are designed to withstand an estimated output of 2,000 hp. (Photo Courtesy Racing Head Service)

BMP LS Block

The Bill Mitchell Products (BMP) LS7X race block is an aluminum four-bolt steel main cap block that features an OEM main bore (to accept any LS crank). Deck height choices include standard LS 9.240 inches or 9.800 inches. Cylinder bores are available in 4.000 inches and can accept up to 4.185 inches. Other features include priority main oiling, LS and Gen I motor mount locations, billet-steel main caps, cam bores that can be machined to 60 mm, and six-bolt-per-cylinder-head bolt locations. The block will readily accept crank strokes up to 4.250 inches.

The BMP LS block is cast aluminum and will accommodate up to a 4.185-inch bore and 4.250-inch stroke. The block has many unique features that support max-performance builds. The main caps are 1045 steel alloy rather than 1020. The block has priority oiling that lubricates the crankshaft mains first and the top end second. (Photo Courtesy BMP)

Crankshafts and Connecting Rods

All Gen III and Gen IV crankshafts share a similar design, but construction materials, stroke length, and engine oiling system requirements differ among versions. Most OEM LS crankshafts are made of cast iron, except for stronger forged cranks used in the LS7 and LS9 engines. LS2, LS3, and LS9 cranks feature a 3.62-inch standard stroke, while the LS7 crank features a 4.00-inch stroke. To accommodate the oil pump drive on LS7 and LS9 engines, the crankshafts feature a snout that's about 1 inch longer than other LS crank snouts. What remain the same among all LS cranks are main journal diameters of 2.65 inches and rod journal diameters of 2.10 inches. All LS cranks also accept the same one-piece rear main seal. All rod journals have the same spacing to accommodate the block's 4.400-inch bore spacing, and all LS blocks feature the same cylinder bore spacing.

The crank should *always* be treated to a balancing job when swapping an LS crank from one version to another, regardless of what pistons and rods are to be used, and whether

Crankshaft Specifications

Crankshaft Stroke	
6.0L LS2 (364 ci)	3.62 inches
6.2L LS3/LS9 (376 ci)	3.62 inches
7.0L LS7 (427 ci)	4.00 inches

Crank Flange	
2006–2008 LS2	Six-bolt
2009 and later LS	Eight-bolt
2009 and later LS9	Nine-bolt

Crank Journals	
Main journal diameter	2.650 inches
Rod journal diameter	2.100 inches

GM Crankshaft Part Numbers		
Type	PN	Cast/Forged
LS2	12588612	Cast iron
LS7	12611649	Forged steel
LS3	12597569	Cast iron
LS9	12603616	Forged steel

Crankshaft Position Sensor		
Gen III	Black (for 24-tooth reluctor wheel)	PN 12560228
Gen IV	Gray (for 58-tooth reluctor wheel)*	PN 12585546
*The LS9 sensor is specific to the LS9, as PN 12601389.		

All LS crankshafts are cast iron except for the LS7 and LS9 forged-steel units. The cast cranks feature a main journal diameter of 2.650 inches and a rod journal diameter of 2.100 inches.

While GM OEM cranks accept the crank pulley as a press-on fit with no indexing keyway, aftermarket performance cranks usually do include a front key to accept a keyed crank balancer. If you plan to run a keyed aftermarket balancer such as ATI or Fluidampr, make sure that the crank has a front keyway.

Both OEM and aftermarket crank snouts feature a key to locate the oil pump drive gear. Aftermarket cranks are available with or without a damper key. If you plan to run a keyed aftermarket damper, the crank snout must feature a front key.

All Gen IV cranks feature a 58-tooth reluctor wheel. When purchasing a new crank, make sure that the reluctor wheel is already installed. If you buy a crank and the wheel is separate, a special indexing tool is required to position the wheel in the correct clock position. Goodson Tools & Supplies offers such a tool. The wheel is carefully pressed into position. Never hit or tap the wheel, which is easily distorted. The wheel should never move if properly interference-fit. Some builders prefer to add a tack weld to secure it from potential rotation.

If a tone wheel is damaged, or if you purchase a crankshaft that was shipped without the tone wheel already installed, a special tone wheel indexing tool is essential.

using GM or aftermarket pistons and rods. Never assume that the crank is accurately balanced, even if you intend to use rods and pistons that were originally part numbered for a specific crankshaft. Spending $150 to $250 or so for a balancing job is well worth the investment to verify correct balance. An out-of-balance crankshaft can quickly result in catastrophic damage to the crank, rods, pistons, bearings, and potentially even the block.

Crankshaft Reluctor Wheel

The crankshaft reluctor wheel, often referred to as a "tone" wheel, features a precision-spaced number of teeth along the circumference, which provides a timing reference for the ECM. As the crankshaft rotates, signals produced as the reluctor wheel passes by the stationary crankshaft position sensor relay the crankshaft's position to the ECM. In combination with the camshaft position sensor, these signals provide the ECM with information used to regulate ignition spark timing and fuel injector timing. The reluctor wheel is interference-fit to the rear crankshaft flange.

The tooth count of the reluctor wheels began with 24-tooth versions and evolved to the use of 58-tooth versions. The greater number of teeth provide a slightly more precise timing signal, allowing the ECM to control spark more. To provide a relative example, think of it as a desktop computer that has a stronger processor for faster and more accurate processing.

In practical terms, it really doesn't matter which tooth count you have, as long as the ECM is compatible and designed to work with that specific tooth count. For instance, if you're buying an aftermarket ignition controller for an LS application, you simply need to choose the 24-tooth or 58-tooth controller, based on the tooth count of your crankshaft's reluctor wheel. If you have a 24-tooth wheel and for some reason you wish to change to a 58-tooth wheel, the reluctor wheel may be swapped out. Be aware that a special indexing tool is necessary when installing a wheel to the crank, to properly index the wheel. This special tool is available

A special tool fixture is required to install a reluctor/tone wheel to an LS crank. The Goodson Tools & Supplies reluctor wheel installation tool, PN RRJ-350 Reluctor Ring Jig, will work with either the 24- or 58-tooth wheel (24-tooth wheel seen here).

The installation tool features an 8-mm locating dowel that engages into the tone wheel and is located on a welded tang on the outside of the tool body.

Inside the tool body is an internal dowel that locates into the crankshaft's rear flange dowel hole.

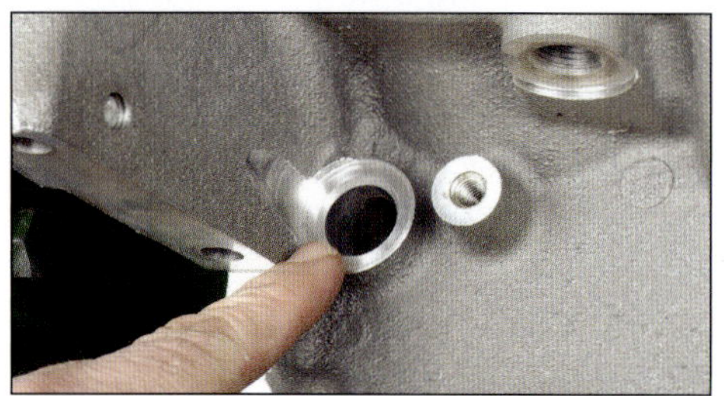

The crankshaft position sensor mounts to a smooth-bore hole on the right rearward side of the block. This hole aligns to the teeth of the reluctor wheel. Gen III sensors are black in color, while Gen IV sensors are gray. Gen III sensors are suited to 24-tooth tone wheels, while Gen IV sensors match to 58-tooth wheels.

The registering tool's internal locating pin provides a high-precision fit into the hole in the crank flange.

Note how the installation tool's outer dowel pin precisely engages into the tone wheel's 8-mm indexing hole.

With the wheel heated and affixed to the tool, the wheel will easily slip onto the crank flange. Note that the tone wheel is indexed on the crank with the two arrowhead-shaped holes at 90 degrees to the first rod pin, with the series of large holes in the wheel positioned opposite from the first rod pin. However, you simply cannot install the wheel by eyesight alone. The specialty tool is mandatory to properly locate the clock position of the wheel.

from suppliers such as Goodson Tools & Supplies.

Some early Gen IV engines feature a 24-tooth crankshaft reluctor wheel (2005 LS2 Corvette, 2005–2006 LS2 GTO, 2005–2006 LS2 SSR), GM part number 12551520 or 12559353. Later Gen IV engines feature a 58-tooth reluctor wheel (2006–2007 LS2 Corvette, 2006–2009 LS7 Corvette, 2008–2009 LS3 Corvette, etc.), as GM PN 12586768.

Note: The LS9 crankshaft reluctor wheel also features 58 teeth, but it is slightly smaller in diameter and requires a different crank position sensor. As of this writing, General Motors apparently does not have the specific reluctor wheel for the LS9 available as a service part. However, the LS9 crankshaft position sensor is available as PN 12601389.

Crank Snout Length

Dry sump engines such as LS7 and LS9 feature longer snouts, approximately 1 inch longer than other LS crank snouts, to accommodate the dry sump drive.

Standard (wet sump) crank snout length is 3.3880 inches. LS7 and LS9 dry sump crank snout length is 4.2740 inches. Dry sump and supercharged applications require a longer crankshaft snout to accommodate the dry sump oil pump drive.

Note: Main crank journal and main bearings are the same size for all LS engines. The only difference with the GM main bearings for LS7 and LS9 applications is the addition of an antifriction bearing coating. If buying aftermarket main bearings (in standard size), an example is the Clevite

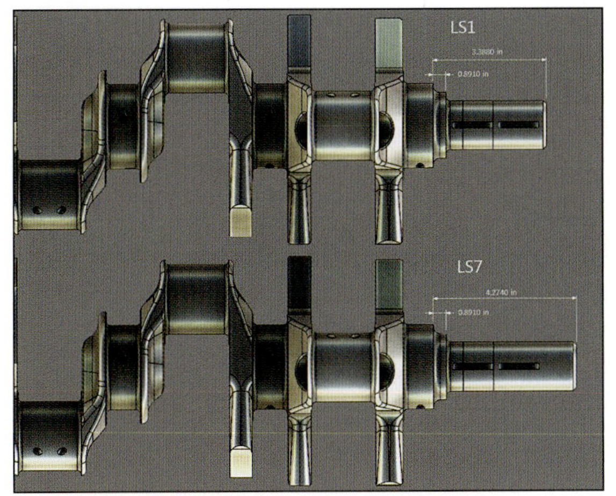

Here's a comparison of crankshaft snout length between LS1/LS6/LS2/LS3 cranks and LS7/LS9 cranks. The snout length of most LS cranks is 3.3880 inches, while the snout on LS7 and LS9 cranks is longer, at 4.2740 inches, to accommodate the dry sump drive. The bevel for the crank gear is identical on all LS cranks, at .8910 inch deep. (Photo Courtesy Callies Crankshafts)

GM Bearing Part Numbers

Main Bearings	GM Part Number
All except LS7/LS9	88894271
LS7/LS9	89017877
Main thrust (except LS7/LS9)	89017572
Main thrust LS7/LS9	89017808

MS-2199HK. These bearings are available coated.

Connecting Rods

All OEM connecting rods for LS engines are made of powdered metal except the rods for LS7 and LS9 engines, which are forged titanium. Powdered metal rods are surprisingly strong and are generally acceptable for horsepower applications up to around 400 to 450 hp. For higher-horsepower assemblies, it is strongly recommended to upgrade to quality aftermarket forged-steel rods, such as those offered by Scat, Eagle, Callies, Lunati, Manley, and others.

Rod-to-piston design changed as well from earlier models. Pre-2006 rods featured a press-pin design, while 2006 and later feature a floating pin design that has bronze bushings installed in the rod small end and retaining clips to secure the wrist pins.

The LS7 rod small-end front and rear outer surfaces are machined at a partial relief cut angle to accommodate the LS7 piston pin boss design, so you can't use OEM LS7 pistons with rods from a different version. Another unique aspect of the OEM LS7 and LS9 titanium rods involves the rod big-end bore diameter, which differs from other LS rods. Although the crankshaft rod journals are the same diameter as other LS cranks, the LS7 and LS9 rod bearings require GM PN 89017811, as opposed to LS1/LS2 rod bearings under GM PN 89017573 (or equivalent crossover to aftermarket bearings).

Keep in mind that we're talking about OEM parts. When dealing with performance aftermarket rods, rod bearing OD may be identical regardless of the engine version.

Rod Bearings

Aftermarket rod bearings are readily available. An example is the Clevite CB-663HNK (standard size). Rod bearings are offered in a variety of under and over sizes. CB-663HX bearings, for example, are slightly thinner, allowing an extra .001 inch of oil

While most LS engines feature powdered metal connecting rods, a performance upgrade should include a set of forged-steel H-beam rods. The factory PM rods are actually very strong, but if horsepower expectations exceed around 450 hp, upgrading to forged-steel rods is recommended.

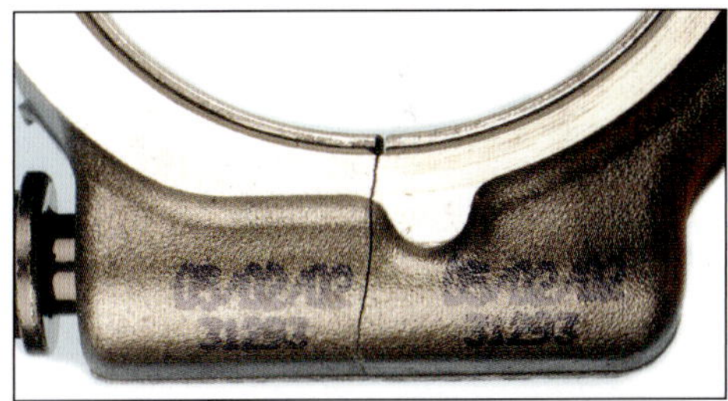

Powdered metal rods (known as PM rods) are pressure cast as one piece. The parting line is then created by snapping off the cap. This creates an uneven and unique mating surface for each rod and cap. When assembled, the cap registers perfectly to the rod because of its unique "fingerprint."

PM rods are also known as "cracked cap" rods for an obvious reason: the method involved in separating the rod cap. The interlocking nature of the cap-to-rod mating provides a very accurate alignment.

clearance (or when mixed with a standard bearing on the same journal, this provides an added .0005 inch of oil clearance). Under sizes for reground crank journals are also available.

As mentioned earlier, factory original LS7 and LS9 applications utilize a different thickness of rod bearing due to the unique LS7 and LS9 rod big-end diameter.

Several respected performance aftermarket manufacturers offer upgraded crankshafts and connecting rods for the LS family of engines.

High-quality, high-strength aftermarket rod bolts such as those made by ARP are not torque-to-yield and can be reused. This eliminates the need to perform a torque-plus-angle tightening, as these are designed for a straight torque application. Also, each end of the rod bolt (shank tip and head) features a centered dimple that allows the use of a rod bolt stretch gauge to monitor bolt stretch during installation for a more precise clamping load.

OEM and aftermarket full-float piston pin applications feature a bronze bushing. A full-float pin design reduces friction, making it applicable regardless of horsepower level. However, in my opinion, if you plan a build to exceed around 400 hp, a full-float wrist pin system should be considered mandatory.

All Gen III and Gen IV rod and piston combinations feature powered metal connecting rods and hypereutectic cast pistons except the LS7, which features titanium rods and hypereutectic pistons, and the LS9, which features titanium rods and forged-alloy pistons. Forged pistons are required to withstand the higher cylinder pressures caused by supercharging. Shown here is a powdered metal rod and hypereutectic piston. Piston wrist pins are interference-fit to the pistons except for the LS7 and LS9 applications, which feature a full-floating pin.

A performance upgrade of aftermarket rods and pistons will benefit any Gen III or Gen IV build. Various rod lengths and piston compression heights are available to suit any bore and stroke combination. While OEM powdered metal rods and hypereutectic pistons are reasonably durable, if you plan to exceed 450 hp, upgrading to forged rods and pistons is highly recommended. The same applies to forced induction. If you plan to apply serious boost, or if you plan to use nitrous oxide injection, moving to these stronger components is critical.

Aftermarket performance connecting rods feature laser-etched numbers on one side of the big end. If rods and rod caps happen to be mixed up during any rod service, during assembly these matching cap-to-rod numbers allow you to verify the correct cap for any specific rod. Also, since the numbers appear only on one side, this helps to identify cap orientation to the rod. Of course, one side of the big-end bore will feature a larger chamfer, which always faces the crank journal fillet, so the chamfer of the rod saddle and cap must be identical when assembled.

The LS7 OEM connecting rods are made of forged titanium, primarily due to their lighter weight. Note the beveled area of the rod small end, which is designed to accommodate the pin boss bracing of the LS7 OEM piston. Unlike other LS series powdered metal rods that feature a cracked-cap design, where the cap is separated from the rod to provide a perfect mating match-up with no machining required, the LS7 titanium rods feature traditional machined surfaces that are honed to round.

Rod Length

One big difference among Gen IV LS rods involves rod length (center of the wrist pin bore to center of the big-end bore). The 6.0L (LS2) and 6.2L (LS3 and LS9) engines originally used a rod length of 6.098 inches, while the 7.0L (LS7) rod length is 6.067 inches. If moving to a longer crankshaft stroke, rod length will vary to work with the combination of block deck height, stroke, rod length, and piston compression distance (piston CD). Depending on crankshaft stroke and piston CD, listed here are minimum rod lengths based on stroke. Given a stock block deck height of 9.240 inches, the remaining variable will be piston CD:

Stroke	Minimum Rod Length
4.000 inches	6.100 inches
4.125 inches	6.125 inches
4.250 inches	6.300 inches
4.450 inches	6.300 inches
4.600 inches	6.300 inches

Enhancements over OEM components include high-density forgings; a wide selection of crank strokes and rod lengths; choices of I-beam or H-beam rod designs equipped with high-strength rod bolts that are installed by torque rather than torque-plus-degree tightening; lighter-weight rotating assemblies; and more. Offerings in terms of stroke length and rod length allow the builder to achieve the stroke desired for an application. Rods for LS cranks feature on-center bores with no offset to accommodate LS rod pin spacing.

Crank Stroke and Rod Length

If cylinder bore diameter and crankshaft stroke are equal, the engine is referred to as being "square," with a stroke/bore ratio of 1:1. If the cylinder bore diameter is greater than stroke, this is called "oversquare." If stroke is greater than cylinder bore diameter, this is called "undersquare." With an undersquare combination, by increasing the crankshaft stroke, peak torque is created at a lower RPM range and piston speed increases at higher RPM. By increasing bore diameter and/or stroke, you increase engine displacement.

Determining the proper crank stroke, rod length, and piston compression distance directly affects the piston deck height in a specific block. For example, an LS block with a deck height of 9.240 inches (crankshaft bore centerline to head deck surface) requires a crank stroke, rod length, and piston pin height combination that will achieve an assembled deck height that does not exceed the block's deck height; otherwise the piston would be above block deck at TDC. Most OEM LS blocks feature an assembled deck height that would bring the piston dome to .005 to .010 inch below the block deck. After machining to square the block decks to correct for factory tolerances, it's not uncommon for assembled deck height to bring the piston .005 to zero the deck height.

As an example, we'll pose a zero-deck assembly for the sake of illustration. When considering assembled deck height, we need to factor only half of the stroke because we only want to consider the distance/length from the crank's rod journal at TDC to the block deck.

Here's a simple formula:

(Stroke Divided by 2) + Rod Length + Piston Pin Height = Assembled Deck Height

Using a 4.000-inch-stroke crankshaft with 6.125-inch rods and piston pin height at 1.115 inches:

(4.000 Divided by 2) + 6.125 + 1.115 = 9.240 inches

If the required block deck height is 9.240 inches, the example package that includes a 4.000-inch stroke, 6.125-inch rods, and 1.115-inch piston pin height would result in a theoretical assembled deck height of 9.240 inches, which would result in a zero deck.

If you decide to increase crankshaft stroke, clearances must be checked as mentioned earlier, including crank counterweight to block pan rail, piston skirt to counterweight, rod big end to block pan rail and bottom of bores, and rod big end to camshaft lobes.

Aside from clearance issues, we need to understand the effect of connecting rod ratio, which involves the leverage effect of the rod relative to the crankshaft. Rod ratio is represented by rod length divided by crank stroke.

An example of a stroker combination that features 6.125-inch rods and a 4.000-inch stroke would be a rod ratio of 1.53:1. Compare this to an OEM LS3 package that features a stroke of 3.62 inches and rod length of 6.098 inches, which has a rod ratio of 1.68:1.

As rod ratio is lowered, rod side-loading increases. Side-loading can be understood by imagining that the piston is being shoved against the thrust side of the cylinder wall in addition to traveling up and down through the cylinder. Increased side-loading places additional friction between the piston skirt and cylinder wall. But maintaining a rod ratio within reasonable limits of about 1.44:1 shouldn't be a concern as long as proper cylinder wall clearance, cylinder wall finish, the use of stronger aftermarket forged pistons, and recommended piston rings are adhered to. Granted, certain race-only packages may call for shorter stroke and higher rod ratio, but for street or street/race setups, you should be fine sticking within the stroke and rod length combination limits recommended and offered by performance aftermarket manufacturers. In short, if you want more torque at lower RPM, the bigger the bore and longer the stroke, the better.

Determining Displacement Based on Stroke and Bore Diameter

It's easy to determine theoretical engine displacement with one of two simple formulas. The first formula uses .7854, a factor of pi:

Bore x Bore x Total Stroke x .7854 x Number of Cylinders

Example using a 4.125-inch bore diameter and a 4.000-inch stroke in a V-8 engine would look like:

4.125 x 4.125 x 4.000 x .7854 x 8 = 427.65 ci

An alternative formula for a V-8 engine using the same bore and stroke is as follows:

Bore x Bore x Total Stroke x 6.2838

4.125 x 4.125 x 4.000 x 6.2838 = 427.69 ci

Aftermarket Connecting Rods

Selecting the most suitable connecting rods for your build is essential. Nearly all stock LS connecting rods are made of powdered cast metal, so when high-horsepower engines are built, owners often upgrade to forged or billet rod in H- or I-beam construction. Titanium rods are installed in the LS7 and the LS9.

The connecting rod is under enormous stress and must support the power output of the engine; therefore, you need to install a connecting rod that meets or exceeds the engine's requirements. OEM powdered metal rods are adequate for medium-HP street builds, but if you plan for more than 400 hp, it is strongly recommended to upgrade to forged-steel connecting rods. When deciding between H-beam or I-beam rods, an I-beam design is slightly lighter than an H-beam rod, making an I-beam perhaps more desirable in terms of higher RPM. However, while an H-beam rod may be slightly heavier, the H-beam design offers greater rigidity and stiffness more suitable for high-torque applications and is able to better withstand high compressive force. Aluminum, billet, and titanium rods offer lighter weight but come at a higher cost. For the majority of high-performance street builds, forged-steel H-beam rods are generally the best choice, considering both durability and budget.

Callies

Callies has a rod selection that includes Ultra XD, Ultra H-Beam, Ultra I-Beam, and Compstar lines in both H- and I-beam construction. There are sizes for 2.100-inch rod journals.

Brian Crower

The Pro Series H-Beam features 4340 steel with radial grooved pin bushings, a lightening hole above the big end, and double-ribbed cap for added strength. Lengths include 6.100 inches and 6.125 inches.

Eagle Specialty Products

Eagle offers a range of H-beam and I-beam rods for LS fitments. Within its LS Series, lengths include 6.100, 6.125, 6.200, 6.250, 6.460, and 6.560 inches, all with .927-inch wrist pins. The standard rod pin size is 2.100 inches, but lengths of 6.460 and 6.560 are also available for 1.8890-inch rod pins. The LS Series for L92 heads includes I-beam design in a 6.125-inch length. The company also offers applications for the LSX tall-deck 9.700-inch block and for the LS World Warhawk 9.800-inch-deck block.

GRP Connecting Rods

GRP specializes in billet-aluminum and billet-titanium rods, made to order. Billet rods are completely CNC-machined to the builder's specifications, so center-to-center lengths and bore diameters can be achieved per the builder's wishes. High-density forgings are CNC-machined to shape. Center-to-center distance is CNC-machined to within .0002 inch. Final bore sizing is also finished to the same tolerance. The rods are then highly polished to remove any potential stress risers. Consider billet-alloy rods only for extreme racing applications.

K1 Technologies

K1 offers a steel-billet H-Beam in either 6.098-inch or 6.125-inch center-to-center lengths. It's difficult to place a horsepower rating on connecting rods. Performance rods are designed, including material of choice, profile, and dimensions, to handle high loads experienced in high-performance and racing applications. Several variables affect the rod's ability to handle high loads, including peak operating RPM, weight of the piston, and crankshaft stroke. Heavier pistons, higher engine speed, and longer crank stroke are all factors that increase inertia loads on the rods. If we're speaking in very broad generalities, high-quality forged rods should be able to withstand up to about 1,000 hp, but again, this is assuming that the rods are not overloaded, that bearing clearances and oil delivery were not compromised, and that rod caps have been properly tightened with the appropriate rod bolts at the correct torque value while not exceeding rod bolt stretch limits.

Lunati

The company's Voodoo Series rods feature H-beam design, while the Signature Series rods feature I-beam. All Lunati LS rods are offered in lengths of 6.125, 6.250, 6.300, and 6.400 inches. All accommodate a 2.100-inch journal size. All rods are CNC-machined forgings of 4340 steel. A number of variables affect the connecting rod's ability to withstand the inertia loads produced by piston weight, length of crank stroke, and engine speed, in addition to engine tuning and piston-to-head and piston-to-valve clearances. When combustion occurs, the piston is pushed down, placing a compressive load on the rod. At TDC on the exhaust stroke, the piston is trying to continue upward while the crankshaft is trying to pull it back down, resulting in tension exerted on the rod. With all of this said, a quality forged connecting rod is expected to handle in the range of 1,000 hp. For big power builds, select high-quality forged or billet rods that have been properly designed, machined, stress relieved, and inspected for anticipated performance applications by any of the established aftermarket rod makers. If you plan to exceed 400 hp, it's best to shy away from powdered metal OEM rods.

Manley

Manley's H-beam and I-beam construction comes in lengths of 6.100 and 6.125 inches. In addition, it offers longer rods for tall deck blocks, including 6.300 inches for RHS tall deck blocks, 6.350 inches for World Warhawk tall deck blocks, and 6.460 inches for GM LSX tall deck blocks. While either H-beam or I-beam forged rods offer far superior strength as opposed to pressure-cast powdered metal rods, H-beam rods are generally considered a bit stronger than I-beam designs.

MGP Connecting Rods

MGP specializes in aluminum rods, custom made to order, for horsepower ranges from 600 to 1,100 hp. Aluminum rods, once considered only for extreme horsepower builds, are becoming more popular for those building upward of 240 to 300 hp per cylinder. Aluminum rods offer a few advantages, including lighter weight for higher revs and the ability to reduce bearing wear, as aluminum tends to absorb shock more than steel rods. However, aluminum rods are larger than steel rods in terms of material thickness, so more attention needs to be paid to rod-to-block and rod-to-cam clearance. Also, aluminum rod length tends to grow more under stress, so an additional .010 inch or so should be added at the piston-to-head clearance. The downside to aluminum rods is expense; aluminum rods are more costly than steel rods.

Oliver

Oliver offers rods in both standard 2.100-inch journals and 2.000-inch journals, at a 6.125-inch length. In addition, its Speedway Series offers rods for blown and turbo applications of up to 2,000 hp.

Scat Enterprises

Scat offers a variety of performance rods in both H-beam and I-beam designs, machined from 4340 steel, including the Premium Pro, Premium Pro Sport, and Ultra Lite Series. A variety of lengths are available, as well as choices of wrist pin diameters of .927 and .945 inch. Scat's stroker rods provide low-profile big-end shoulders for increased clearance.

Venolia

Venolia specializes in custom-order aluminum rods for all-out racing applications. Bear in mind that aluminum rods are thicker than steel rods, so extra attention must be paid when checking big-end clearances to the block pan rail, cylinder bottoms, and other components.

High-Performance Crankshafts

Crankshaft are available in cast, forged, and billet materials. Cast iron or steel is the most economical, as molten metal is poured into a mold, with subsequent machining, heat treating, and stress relieving and nitriding for hardness. Cast cranks are generally heavier and somewhat more brittle compared to a forging, being more prone to fractures under high-stress conditions. However, for the average factory stock street car, with the engine built to a moderate power level, a cast crank will get the job done. It's difficult to give a horsepower rating because many variables enter the picture, such as quality of the material, balancing, assembly technique, and level of abuse, to name a few. If I were to assign a number, a quality cast crank should be acceptable up to about the 500-hp range. A quality aftermarket cast crank runs in the range of about $250 to $450. Today's cast cranks offered by leading manufacturers provide far higher quality and greater durability than in decades prior.

For high-horsepower performance, forged cranks are much stronger and will better withstand the forces applied by the rod and piston. Forged cranks are generally made using one of two metal makeups: 4130 or 4340. Both feature a percentage of molybdenum for added strength, with 4130 having about 20 percent moly and 4340 about 25 percent moly. Both nickel and carbon are part of the mix, to add hardness and grain uniformity. The 4340 cranks feature a greater carbon and nickel content. The 4130 cranks are a bit lighter, while 4340 cranks offer the highest strength. This is a very brief explanation of the material mix numbers, but the point is that 4340 cranks are more suited to high-performance applications. Expect to pay in the neighborhood of $700 to $1,000 for a quality forged crank.

Billet cranks start as high-density bar-stock forgings with a very uniform grain pattern, 100-percent machined to final shape and finish. Top makers of billet cranks choose the highest-grade billet bar stock, taking care to avoid any material that may contain inclusions that can result from hammer forging, where atmospheric particles might be introduced. They go to lengthy procedures to ensure that tensile strength is optimized. Not surprisingly, billet cranks are more costly to manufacture, and they can run $2,000 to $3,000. Billet cranks are commonly used in NASCAR and 10,000-hp Top Fuel engines.

Why the disparity in prices between forged cranks and billet? Producing forged crankshafts requires designing and making a two-piece mold, which is a hefty investment. However, forged cranks for common applications are made in quantity, keeping the price affordable. A billet crank requires no mold and the CNC-machining process allows the maker to machine any profile, so you might first think that it would be less expensive, but this still requires buying much more expensive billet bar stock, and CNC-machine hours to produce "one off" cranks. If you can afford a high-dollar billet crank, there's no downside other than the higher cost. For most of us, a quality forged crank will fill the bill and should handle upward of 2,000 hp.

Some forged cranks are made using a twist or nontwist method. A twisted crank is made with rod throws inline in a single plane. After the initial forging process, specific main journals are then heated to a malleable state. The crank is then twisted to place the rod journals into their required offset positions. This causes the grain of the steel to be interrupted and nonuniform, potentially decreasing crank strength. All the aftermarket crank makers that I'm aware of use a nontwist approach. A nontwist forging is created by forging the crankshaft in two planes from the very start, placing the rod pins in the proper angles, requiring only finish machining, stress relieving, and heat treating.

Manufacturers of high-performance crankshafts offer several enhancements that improve durability and the ability to rev faster and at higher RPM while reducing weight. Examples include gun-drilling, where the main centerline is drilled to remove mass. Reducing weight at the centerline of rotation won't make the crank rev higher, but this is a step that reduces crank weight, and any weight reduction to the vehicle is beneficial. Rod journals are drilled through to reduce both overall and rotational weight. Fillets, the corners of the bearing journals that meet the counterweights, are generously radiused instead of being square cut, which enhances strength by removing stress risers. A feature that may be standard or optional, depending on the crank, is aero-profiling the counterweights, where the trailing faces of the counterweights are tapered much like the trailing edge of an airplane wing. This reduces parasitic oil weight as the crank spins, allowing oil to leave the counterweight surfaces quicker, reducing drag. Another enhancement, not really necessary for most street-strip performance builds, includes removing mass from the counterweights in order to produce a lightweight crankshaft by CNC-carving or skeletonizing the counterweights.

Bryant Racing

Bryant specializes in custom-ordered cranks machined from billet steel, targeted at pro-level competition use. Any sensible stroke, journal diameter, or custom feature is available. Its Ultralight Package offers counterweight grooving, lightening holes, scalloped flange, etc., to produce the lightest crank that will still handle high loads. It also offers in-house REM Isotropic super finish, which produces a low Ra finish for enhanced friction reduction.

Callies

Callies offers an extremely wide selection of cranks, including Ultra Billet, Magnum, Magnum XL, Stock Eliminator, and Dragonslayer models. Designs range from stock configuration to super-exotic scalloped and lightened versions for street to street/race to all-out competition.

Eagle Specialty Products

Eagle's line of 4340 forged-steel cranks for LS applications includes stroke choices of 3.6220, 4.000, 4.125, 4.250, and 4.3750 inches, handling horsepower from 1,000 to 1,500. All are available with either 24- or 58-tooth reluctors.

K1 Technologies

All cranks are forged steel with 2.650-inch main and 2.100-inch rod journals. Available strokes (each with either 24- or 58-tooth reluctors) include 3.622, 3.900, 4.000, 4.100, 4.125, and 4.250 inches.

Kings Crankshaft

Kings is another firm that specializes in custom billet crankshafts, made to order. Starting with high-density forged-billet blanks, the crankshafts are then entirely precision CNC-machined.

Lunati

All Lunati cranks are made of 4340 nontwist forged steel. Three different lines of cranks are available: the Voodoo Series, Signature Series, and Signature Series Blower. All feature 2.559-inch main and 2.100-inch rod journals.

The Voodoo Series' (non-LS7 type) available strokes include 4.000 or 4.100 inches.

The Signature Series is Lunati's premium performance line, with rated capability of handling more than 1,500 hp. Features include gun-drilled mains, micro-polished journals, windage-reducing contoured wing counterweights, and plasma nitriding. Strokes available in the non-LS7 type include 3.622, 4.000, 4.125, 4.250, 4.500, and 4.600 inches. The LS7 long-nose line offers strokes of 4.000, 4.125, 4.500, and 4.600 inches.

The Signature Series Blower line, intended for high-pressure forced induction, includes dual keyways, larger-diameter nose bolt threads, and larger-diameter flexplate flange threads. The Blower line offers the non-LS7 type in strokes of 4.000, 4.250, 4.425, 4.500, and 4.600 inches. The Blower long-nose LS7 type is available with strokes of 4.000, 4.125, 4.250, 4.500, and 4.600 inches.

Manley Performance Products

Manley offers three lines of cranks: Lightweight LS non-LS7, Lightweight LS7, and Super Lightweight. All are machined from 4340 forgings, all with 2.559-inch main and 2.100-inch rod journals. Note: Manley cranks are not intended for highly boosted Roots-style cog belt supercharger applications.

The Lightweight LS (standard snout length) is offered in strokes of 3.622, 4.000, and 4.100 inches, available with either 24- or 58-tooth reluctors. The Lightweight LS7 type (long nose) is available in strokes of 4.000 or 4.100 inches, with a 58-tooth reluctor only. The Super Lightweight line is available in a 4.000-inch stroke with 24- or 58-tooth reluctor. Features include nitriding, stress relief, gun-drilled mains, lightened rod journals, and fully profiled counterweights.

Scat Enterprises

Scat cranks are made of 4340 forged steel that is heat treated, shot-peened, and nitrided for added hardness. All LS cranks feature 2.100-inch rod journals, lightened rod throws, gun-drilled mains, large radius on all journal fillets, and dual linear post keyways for keyed dampers. Available strokes include 3.622, 4.000, 4.125, and 4.250 inches.

Winberg Crankshafts

Winberg cranks are intended for upper-level professional racing applications, and they are primarily custom ordered for stroke and journal size. All are fully counterbalanced and are available with or without reluctors. LS1-style cranks (applicable to all blocks except LS7 and LS9) are offered in strokes of 3.250, 3.400, 3.500, 3.625, 3.750, and 4.000 inches, with rod journals of 2.000, 2.100, or 1.888 inches. The long-snout LS7/LS9 cranks are offered in 3.400-inch or 4.000-inch strokes. All feature a main diameter of 2.559 inches.

LS7, LS3, and LS9 Cylinder Heads

Earlier Gen III LS heads, such as those found on LS1, LS6, and various Q Series truck engines, feature tall intake ports referred to as "cathedral" ports. The cathedral-roof shape at the top accommodates the fuel-injector path. The LS2 engine also features cathedral-port heads but is technically considered a Gen IV engine due to the change in cam sensor location. We'll limit our discussion to Gen IV heads with rectangular intake ports.

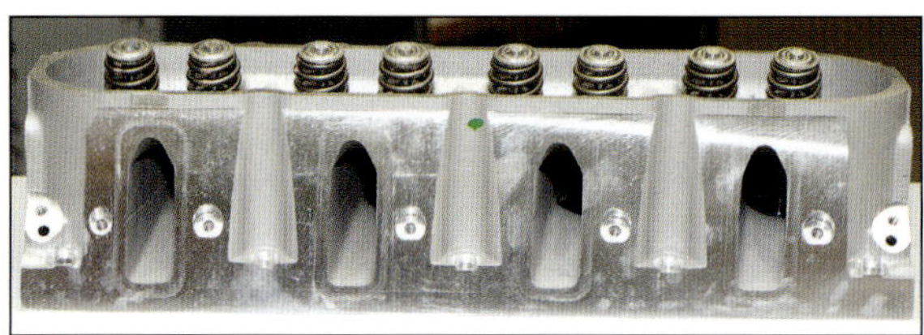

All Gen III heads feature tall and narrow cathedral intake ports. This includes LS1, LS6, and LQ truck variants. The LS2 also features cathedral-port heads, although it's considered a Gen IV engine primarily because the cam timing sensor was moved from the rear of the block to the front timing cover. The cathedral-port design is generally noted for better low-end torque, while the Gen IV rectangular ports are generally known for superior midrange and top-end performance. Cathedral ports measure about 1 inch in width and about 3.125 inches in overall height.

In the Gen IV versions of this later generation of LS engines, we're dealing with two separate cylinder head castings. The LS3, LS9, and L92 all use essentially the same cylinder head foundation, featuring a four-digit identification number, which is at the top of the head just outside the valve cover rail. The casting number for the LS3 head is 0821. The LS9 head features no such casting number; rather, the designation "LS9" is lightly engraved on the lower left of the exhaust side of the head. The LS7 cylinder head is unique, with identification number 8452.

Gen IV "rectangular"-port cylinder heads include the LS7, LS3, L92, and LS9 variants. Basic differences include intake port dimensions, rocker arm mounting pedestals, and combustion chamber volumes.

The LS3, LS9, and L92 heads are "as-cast" heads, with intake and exhaust ports and combustion chambers shaped during the precision casting process. The LS7 cylinder head features intake and exhaust ports and combustion chambers with a more refined dimensional and surface finish that is CNC-machined.

Intake port dimensions also vary between the LS3, LS9, and L92 and the LS7 head versions. The intake ports on the head used for the LS3, LS9, and L92 are 1.250 inches wide x 2.550 inches tall, while the intake ports on the LS7 head are 1.350 inches wide x 2.40 inches tall.

The LS7 cylinder head was

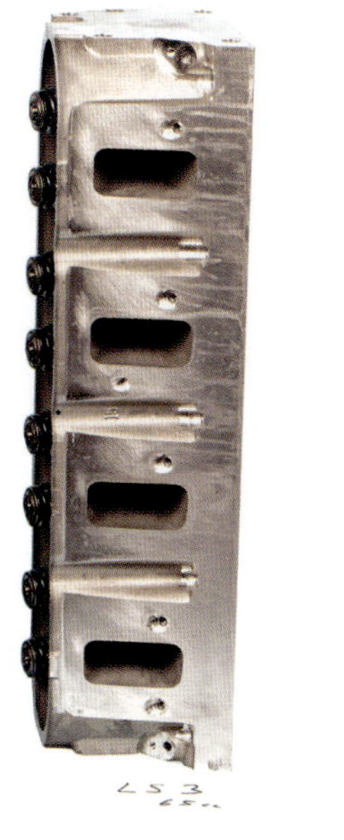

From left to right: LS7, LS3, and LS9 cylinder heads. Notice the taller intake ports on the LS3 and LS9 heads.

Shown left to right: LS7, LS3, and LS9 heads. Combustion chamber volumes differ among the three variants. The LS7, L92, and LS9 feature 70-cc chambers, while the LS3 has 68.4-cc chambers.

Shown here is a cutaway view of an L92 head. The intake port runner design differs slightly among the various head versions, but all are very similar in shape.

offered in 2007–2009 Corvette Z06 models. Identifying an LS7 head versus the LS3, LS9, and L92 heads is fairly easy. The LS3, LS9, and L92 heads feature flat rocker pedestals (to accept separate rocker arm rails), and the ports and chambers feature a cast finish. The LS7 head features individual rocker arm radiused stands and all ports and chambers display a machined surface.

The LS3 cylinder head has become very popular, in part because of the improved flow of the Gen IV rectangular-port design and its relatively attractive street price. General Motors made a lot of LS3 engines along with its truck counterpart, the L92, and as a result these heads are available in great numbers in salvage yards. Even brand-new head prices have become very reasonable; hence their popularity.

Note that while stock cylinder heads feature a four-bolt-per-cylinder-head bolt layout, GM LSX versions are offered with a six-bolt-per-cylinder layout. Aftermarket heads are also available in four- or six-bolt versions.

LS7

LS7 cylinder heads feature a larger combustion chamber and wider valve layout. Because of this, LS7 heads cannot be installed onto blocks that feature a cylinder bore size of under 4.100 inches, as valves may contact the top edges of the cylinder bores. LS7 heads may only be installed onto blocks that feature a 4.100-inch or larger cylinder bore diameter. Unlike

LS7 heads are casting number 8452 and feature slightly raised ports, altered valve angle, and larger titanium intake valves. Exhaust valves are sodium filled and have been known to cause failures. A popular remedy is a change to stainless-steel exhaust valves. Approximate flow bench results at .600-inch lift are about 370-cfm intake and about 240-cfm exhaust.

LS7 intake port deck view. Note the shorter intake port height compared to LS3, LS9, and L92.

LS7 Head Specs

PN 12578449 (complete)
PN 12578450 (bare)
Casting number 8452
356-T6 aluminum
Fully CNC-machined chambers and ports
12-degree valve angle
70-cc combustion chambers
270-cc intake volume
85-cc exhaust volume
11:1 compression ratio with stock pistons
Offset intake rockers
1.8:1 rocker arm ratio
Unique intake manifold bolt pattern
2.200-inch titanium intake valves
1.610-inch sodium-filled exhaust valves
Fits 4.100-inch bores or larger
(LS7 heads are also available in various combustion chamber volumes)
Number of 11-mm head bolts: 4 per cylinder/10 total

The LS3 cylinder head has 260-cc intake runners and 92-cc exhaust ports. In addition, it has a combustion chamber volume of about 68.4 cc. These heads fit any LS engine with cylinder bores of 4.000 inches or larger. It also features 2.165-inch hollow-stem intake valves and 1.59-inch solid-stem exhaust valves.

LS3 Head Specs

PN 12629063
Casting number 0821
319-T5 aluminum
As-cast ports and chambers
D-shaped exhaust ports
15-degree valve angle
260-cc intake volume
90-cc exhaust volume
68.4-cc combustion chambers
10.7:1 compression ratio with stock pistons
2.165-inch hollow-stem stainless-steel intake valves
1.59-inch sodium-filled exhaust valves
Fits 4.000-inch bores or larger
Number of 11-mm head bolts: 4 per cylinder/10 total

Shown here is an LS3 combustion chamber with 68.4 cc of volume. All factory production LS heads feature an as-cast combustion chamber except for the LS7, which has CNC-machined chambers.

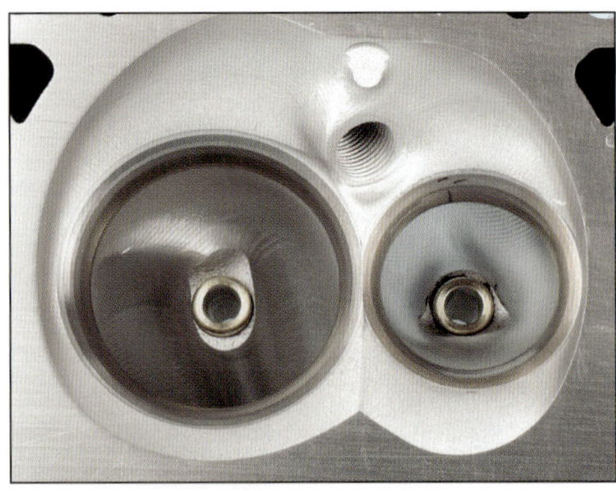

The LS7 heads feature fully CNC-machined combustion chambers with 70 cc of volume.

other factory LS heads, LS7 factory heads are fully CNC-machined at the combustion chambers and ports. Beneficial porting work may involve only port matching the intake ports and/or the intake manifold ports to remove any potential flow obstructions. Due to the LS7 unique intake manifold deck, only intake manifolds designed for the LS7 heads may be utilized due to port match and manifold bolt pattern.

LS3

The LS3 factory head features lightweight hollow steel intake and solid-stem exhaust valves. In stock form, flow bench results have shown flow rates of 296 cfm intake and 208 cfm exhaust, at full valve lift of about .600-plus. This head is intended for blocks that feature a minimum cylinder bore diameter of 4.00 inches.

L92

L92 heads are essentially the same casting as the LS3 heads. The difference is the weight of the intake valves: L92 heads feature solid valves and the LS3 has lighter, hollow-stem valves. L92 and LS3 heads flow pretty well in stock form, but they can ben-

The L92 combustion chamber features 70 cc of volume.

L92 Head Specs

PN 12629064	90-cc exhaust volume
319-T5 aluminum	2.165-inch solid-stem intake valves
As-cast ports and chambers	1.590-inch solid-stem exhaust valves
D-shaped exhaust ports	Fits 4.000-inch bores or larger
15-degree valve angle	Number of 11-mm head bolts: 4 per cylinder/10 total
70-cc combustion chambers	
260-cc intake volume	

efit from porting by opening up the throats a bit and reducing the big rocker bolt bosses in the runners and massaging the exhaust ports. This is best done on CNC for more precise and repeatable results. CNC porting also requires much less time as opposed to hand porting. Performing a good valve job is more critical, making sure that the seats are concentric and that the valves are sunk into the seat at the same depth.

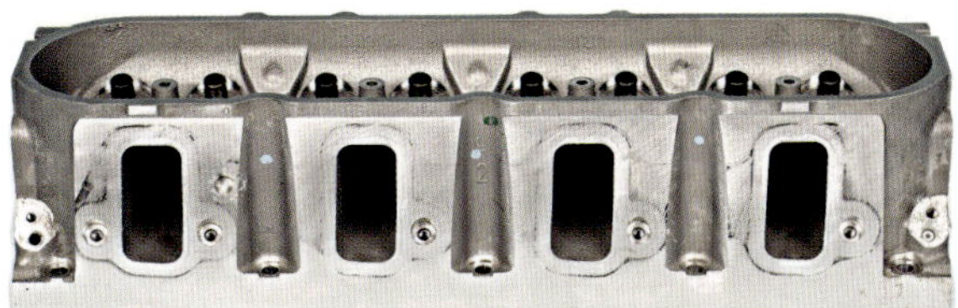

The LS9 CNC-ported cylinder heads have enhanced webbing and a stronger deck for greater rigidity and minimal distortion. The A356T6 alloy also withstands higher-performance service, and in stock trim it is supercharged.

LS9

The factory LS9 cylinder head was designed for higher dynamic compression to suit a supercharger application. This head features a stronger, more-dense alloy along with beefier webbing reinforcements, again to better handle forced induction pressures. Like the LS7 head, the LS9 features a lightweight titanium intake and hollow sodium-filled exhaust valves. To promote air/fuel mixture efficiency, a "swirl wing" is designed into the intake valve boss at the base of each intake runner. The valve guide boss casting features a pronounced vertical riser, or "wing," that promotes a swirl effect before the air/fuel charge reaches the valve. In stock form, full-lift flow has been documented at 270 cfm at the intake and 201 cfm at the exhaust ports. CNC port machining that removes the swirl wing has been reported to increase flow to 322 cfm intake and 218 cfm exhaust.

LS9 cylinder head castings are factory etched with an "LS9" designation on the lower left of the exhaust side, under the spark plug port.

The LS3, LS9, and L92 all use essentially the same cylinder head foundation, featuring the four-digit identification number (top of head just outside the valve cover rail) 5364. The LS7 cylinder head is unique, with identification number 8452.

The LS9 and L92 heads feature flat rocker pedestals (to accept separate rocker arm rails). The ports and chambers feature a cast finish. The

All factory LS heads, including Gen III and Gen IV, are equipped with tapered, or "beehive," valve springs instead of traditional parallel-wound springs. The beehive spring design, which features an oval wire instead of a round wire, features progressively smaller coils from bottom to top. This was intended to reduce retainer mass and diminish valve spring harmonics/frequencies. However, opinions among performance engine builders vary, opting to choose either beehive- or parallel-wound-style springs. Installed height on LS7 heads is 1.960 inches with 310 pounds at 1.370 inches open pressure. LS3 and LS9 springs feature an installed height of 1.800 inches and 1.250 inches at 295 pounds.

LS7 head features individual rocker arm radiused stands and all ports and chambers display a machined surface.

In addition, the LS3, LS9, and L92 heads are originally equipped with the tapered "beehive" valve springs, while the LS7 head features "straight" valve springs without the beehive taper.

An LS9 combustion chamber with 70 cc of combustion chamber volume. The heads come with production lightweight titanium intake valves and sodium-filled exhaust valves.

LS9 cylinder head, deck view. The heads are rotocast from strong A356T6 aluminum alloy, so the molten alloy is equally distributed in the mold for greater density and strength.

Unlike the LS1, LS6, and LS2 heads that feature the tall, skinny cathedral intake ports, the LS3, LS9, L92, and LS7 heads feature a conventional rectangular-shaped intake port. Intake port dimensions also vary between the two head versions. The intake ports on the head used for the LS3, LS9, and L92 are 1.250 inches wide x 2.550 inches tall, while the intake ports on the LS7 head are 1.35 inches wide x 2.40 inches tall.

Cylinder Head Interchangeability

When mixing and matching heads and blocks, it's critical to pay attention to cylinder bore size and the cylinder head's chamber and valve layout. You cannot use a cylinder head that features a valve layout that is too large for the cylinder bore because you'd run the very real risk of valves crashing into the block deck at the cylinder bore edges. For instance, you cannot run an

LS9 Head Specs

PN 12621774
A356-T6 Rotocast aluminum (stronger/more dense)
15-degree valve angle
260-cc intake volume*
89-cc exhaust volume
70-cc combustion chambers
9:1 compression ratio with stock pistons
Supercharged
2.165-inch titanium intake valves
1.590-inch sodium-filled exhaust valves
Number of 12-mm head bolts: 4 per cylinder/10 total**

* PN 19328743 LS9 CNC head features 276-cc intake volume.
** The LS9 head can be installed on LS9 blocks with 12-mm bolts, or on other LS blocks that have a minimum 4.000-inch bore with 11-mm bolts.

Cylinder Head Production Years/Models

LS7
2006–2013 Chevrolet Corvette Z06
2013 Corvette 427 Convertible
2007 Holden CSV GTS
2008 Holden HSV W427
2014–2015 Chevrolet Camaro Z28

LS3
2008–2013 Chevrolet Corvette
2010–2015 Chevrolet Camaro SS (manual only)
2008–2017 Holden Commodore family:
- 2009 Pontiac G8 GXP
- 2009–2013 Vauxhall VXR8
- 2008–2016 HSV Clubsport R8
- 2008–2013 HSV GTS
- 2008–2015 HSV Senator Signature
- 2008–2015 HSV Maloo R8
- 2014–2017 Chevrolet SS
- 2015–2017 Holden Commodore

2008–2016 Holden Caprice family:
- 2008–2016 HSV Grange
- 2015–2017 Holden Caprice V

LS9
2009–2013 Chevrolet Corvette ZR1
2017 Holden HSV GTS-R W1

L92
2007–2013 Cadillac Escalade
2009 Chevrolet Tahoe
2007–2013 GMC Yukon Denali/Denali XL
2007–2013 GMS Sierra Denali
2008–2009 Hummer H2
2009–2013 Chevrolet Silverado 1500
2009–2013 GMC Sierra

Cylinder Head Bolt Note

All head bolt holes in a GM factory block feature 11-mm x 2.0 threads except for the LS9 block, which uses 12-mm x 1.75 threads. Bolt length differs depending on the block. Earlier LS1 and LS6 blocks require two different-length head bolts (11-mm x 2.0 x 100-mm and 11-mm x 2.0 x 155-mm), while 2004 and later blocks, including LS7, LS3, and L92, use same-length 11-mm x 2.0 x 100-mm bolts. LS9 blocks require 12-mm x 1.75 x 100-mm bolts. Aftermarket performance blocks may feature inch-format head bolt hole locations, such as those found on Dart and World Product LS blocks. Two examples are the typical 7/16-14 thread as opposed to 11-mm x 2.0 thread found in all OEM blocks except the LS9, which uses 12-mm x 1.75 bolts. All LS heads also require 8-mm x 1.25 pinch bolts for the inboard bolts at the very top of the block decks. OEM head bolts (excluding the 8-mm x 1.25 pinch bolts) are torque-to-yield (TTY)–style bolts, which require torque-plus-angle tightening and are designed for onetime use. Aftermarket head bolts, such as those offered by ARP, are designed to install using a torque-only method and are reuseable.

Note for those who may not understand bolt sizes: Sizes indicate shank diameter, not the required size of wrench for the bolt's hex head. Experienced builders know this, but some novices tend to get confused. The size of the bolt hex head has nothing to do with bolt size. ■

LS7, LS3, or LS9 head on a block with smaller bores. The LS7 head requires a minimum of a 4.100-inch bore. LS3 and LS9 heads require a minimum of a 4.000-inch bore. An exception is the LS2 block which will accept LS3, LS9, or LQ4 heads.

You need to pay attention to the block's bore diameter. Running a cylinder head intended for a larger bore size can result in valves crashing into the block. For instance, you cannot install an LS3, LS9, or LS7 head on an LS1, LS6, or LS2 block (at least not without some creative deck/bore notching to clear the valves).

While all production-based LS heads and blocks feature a four-bolt design, dedicated performance blocks and heads feature a six-bolt design, where an additional head bolt is added at the inboard and outboard locations, inline with the bore centerline. The added two bolts per cylinder provide extra strength for high cylinder pressures. Blocks that feature this six-bolt design include GM's LSX block, as well as offerings from various performance aftermarket makers, such as Dart.

LS Valves

Valve Type	Material
LS7 intake	Titanium
LS7 exhaust	Hollow, sodium-filled stainless
LS3 intake	Hollow stainless
LS3 exhaust	Solid stainless
LS9 intake	Titanium
LS9 exhaust	Hollow, sodium-filled stainless
L92 intake	Solid stainless
L92 exhaust	Solid stainless

Note: Titanium valves require hardened lash caps at the stem tip.

LS1 and LS6 blocks will accept only LS1, LS6, and LS2 heads.

LS2 blocks can use LS1, LS6, or LS2 heads, as well as L92-style heads, which include LS3 and LS9 heads.

LS3 and LS9 blocks can use LS1, LS6, LS2, LS3, or LS9 heads.

The LS7 blocks can accept any LS head.

All production LS blocks except the LS3 and LS9 feature a 4-bolt-per-cylinder head bolt layout, for a total of 10 head bolts per head. All these heads use 11-mm x 2.0 thread except the LS9, which features 12-mm x 1.75 thread. However, the GM LSX race block, LS3, and LS9, along with a host of aftermarket race blocks, feature a 6-bolt design, with 18 bolts total per head. Two additional bolts per cylinder, one inboard and one outboard of the bore centerline, are added. This provides added head securing and rigidity for high-cylinder-pressure applications.

Regarding head-to-block swaps,

Shown here is an example of a six-bolt head, from Trick Flow. Aftermarket head makers offer LS heads in both four-bolt and six-bolt designs. Six-bolt block and head combinations require a special head bolt or head stud kit and head gasket designed for the six-bolt platform.

A six-bolt head can be mounted to a factory four-bolt block, even though you would not be taking full advantage of the head's six-bolt design. The outer bolt bosses would simply hang out past the block deck. If this is the case, the extra outboard bosses may be milled off if so desired. This would occur only if you happened to get a good deal on a pair of six-bolt heads that you wished to use on your factory block.

the four-bolt heads or the six-bolt heads can be installed on either four-bolt or six-bolt blocks. Installing a four-bolt head to a six-bolt block simply won't provide the added cylinder head clamping available with the six-bolt head. Also, while a six-bolt head can be installed to a production four-bolt block, you simply won't be taking advantage of the added bolt locations offered by six-bolt heads. Appearance-wise, the outboard four-bolt bosses simply hang over the block deck but won't cause an issue. If a six-bolt head is installed on a four-bolt block and you don't like the looks of the extra bosses on the exhaust side of the heads, these bosses can be milled off, but this is not necessary.

Water Temp Holes

All LS heads feature a 12-mm x 1.5 female-threaded hole on the left of the exhaust side, intended for a water temperature sender. The sender is traditionally mounted to the left-hand head, but don't forget to plug the hole on the right-hand head. If you're dealing with a new or used head, this plug will already be in place with any luck. If the plug is missing, you can handle this in one of two ways: purchase a GM plug PN 11610259 for about $5, or simply use a 12-mm x 1.5 bolt with a shank length of about 20 mm along with an aluminum or copper crush washer. If the heads have been reconditioned, this plug on the right head might have been accidentally lost or ignored. It's easy to forget, so make sure to plug this unused water hole; otherwise you'll have a coolant-squirting mess when you start the engine.

Each head features a 12-mm x 1.5 threaded hole on the left area of the exhaust side. These holes are open to the cooling. The water temperature sender installs to the left-side head.

An example of a water hole plug on the right-side cylinder head, using a common 12-mm x 1.5 bolt and crush washer.

Rocker Arm Pedestal Rails

Two different rocker arm pedestal rail versions are used. One rail is designed for use on the LS1, LS6, and LS2. The other is designed for the L92, LS3, L99, and LS9.

The rail for the LS1, LS6, and LS2 features the pedestals centered (height-wise) on the rail. Each edge

Rocker Arm Ratios

LS1	1.7:1 (intake and exhaust rockers identical)
LS6	1.7:1 (intake and exhaust rockers identical)
LS2	1.7:1 (intake and exhaust rockers identical)
L92	1.7:1 (intake rocker offset)
LS3	1.7:1 (intake rocker offset)
L99	1.7:1 (intake rocker offset)
LS9	1.7:1 (intake rocker offset)
LS7	1.8:1 (intake rocker offset)

Gen IV rocker arms for L92, LS3, and LS9 applications feature an offset intake rocker and a straight exhaust rocker, both with 1.7:1 ratio.

Cast-in rocker pedestals are integral to the LS7 heads. The radiused pockets allow rockers to be directly bolted to the heads without the need for a separate rocker mounting rail.

As an example of a pair of performance aftermarket rockers for LS Gen IV applications that require offset intake rockers, this underside view shows the slight offset of the intake rocker (left) at both the pushrod cup and roller valve ends. The offset intake rockers are necessary because of the larger valve diameters used in Gen IV applications.

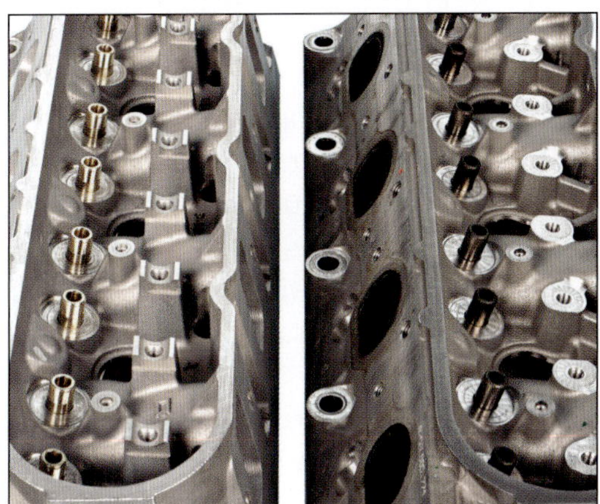

A comparison of the rocker pedestal design on the LS7 head (left) and LS2/LS3/L99/LS9 head versions. Note that the LS7 rocker pedestals are part of the casting, allowing the rocker arms to bolt onto the head with no need for a mounting rail. The other style features flat pedestals, onto which a separate rocker arm mounting rail is installed. The rail is sandwiched between the head and rockers.

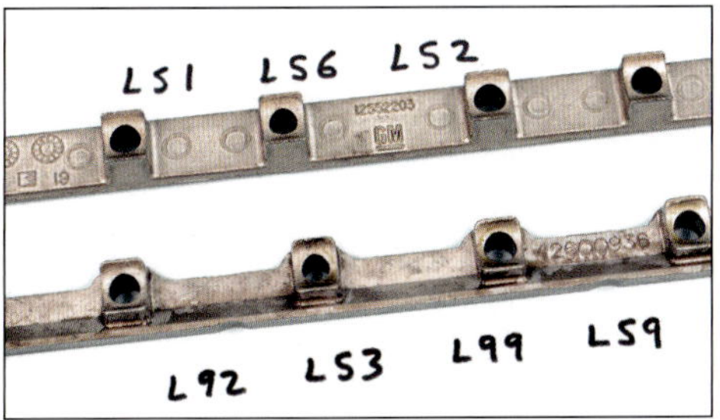

Factory original Gen III and Gen IV heads feature separate rocker mounting rails that allow mounting OEM rockers except for LS7 heads, which do not require a mounting rail. Note that Gen III and Gen IV rails seen here at the bottom are different; the Gen IV rails feature a different rocker arm mount spacing.

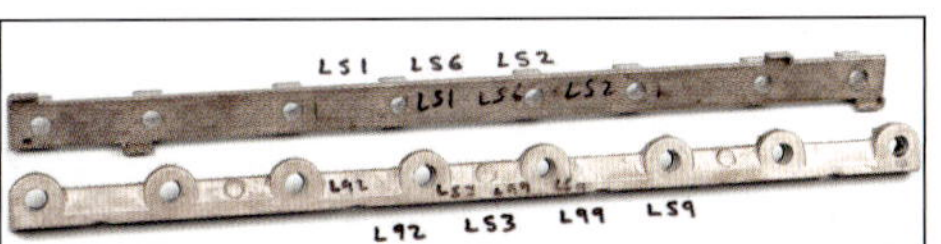

This view shows the underside of the rocker rails. Gen III rails have about the same amount of material above and below the holes, while Gen IV rails feature protrusions with the bolt holes slightly offset from the center length.

of the rail's length is straight. The rail for the L92, LS3, L99, and LS9 locates the pedestals a bit offset, with one side of each pedestal extended out (one edge of the rail features individual pedestal bulges, or radiuses, that protrude out from the edge).

The rail for the LS1, LS6, and LS2 features each cylinder's pair of pedestals located 1.901 inches on center from each other. The rail for the L92, LS3, L99, and LS9 features the pedestal centers located 2.227 inches apart (center of hole to center of hole). The part numbers on the rails are 12552203 for the LS1, LS6, and LS2 rails and 12600936 for the L92, LS3, L99, and LS9 rails. The LS7 cylinder heads feature individual radiused rocker stands as an integral aspect of the casting and do not require the use of a separate rocker arm mounting rail. Rockers bolt directly to the LS7 heads without the need for a separate rail.

Aftermarket Performance Heads

Stock LS cylinder heads provide superior breathing compared to Gen I small-block Chevy heads. Since the introduction of the LS engine, cylinder head architecture has improved, and the original cathedral intake port design has evolved to rectangular-port designs, with the LS3/L92 heads currently leading the pack in terms of performance attributes. Thanks to performance aftermarket manufacturers, even further horsepower gains are to be had. In some cases, gains of roughly 100 hp over stock have been realized because of changes in runner design, valve angle, port height variations, precision CNC-machined chambers and ports, capabilities to accommodate larger/taller aftermarket rockers and springs, increased deck thickness for added rigidity, and more. When you want to extend the power, torque, and RPM envelope, upgrading to aftermarket heads is definitely a move to consider.

Several companies offer aftermarket performance LS heads, including examples from Trick Flow, Dart, Mast Motorsports, Air Flow Research (AFR), Edelbrock, Racing Head Service (RHS), Bill Mitchell Products (BMP), and World Products. A wide array of configurations is available, including those with cathedral-port intakes (Gen III and LS2) as well as those with rectangular-port intakes (LS3, LS7, and L92). Heads are available in bare form and fully assembled, and as-cast or CNC-machined. Choices also abound in terms of intake port volume, combustion chamber volume, and both four-bolt-per-cylinder and six-bolt-per-cylinder formats.

Selecting a Cylinder Head

With all the variations in LS cylinder head platforms, choosing the head that's right for your build can seem daunting at first. The first two considerations involve the type of intake manifold port and the piston bore size of your block. There are three types of factory intake ports: cathedral, square port, and raised square port. The cathedral-port design was featured on Gen III engines LS1, LS6, and LS2, as well as Vortec 4.8L, 5.3L, LQ4, and LQ9 iron blocks. The square port, which is actually a rectangular port, is featured on factory L92, LS3, L99, L76, LS9, and LY6 engines. The raised square port was found on LS7 and GM LSX engines.

All three port designs flow well and produce good power. The cathedral port offers decent torque and is slightly more streetable, with the square-port designs offering higher peak horsepower at higher RPM, more suited for competition use. For the street, cathedral or square is acceptable for moderate power builds. For optimum power at higher engine speed, the square port is preferred. Naturally, the intake manifold must feature the same type of port to match the head.

The piston bore size is a critical factor. The 3.89-inch bore found on LS1 and LS6 engines will accept only cylinder heads that were intended for LS1, LS6, or LS2 applications. Using heads designed for a larger bore size runs the very real risk of valves contacting the edges of the cylinder bore. LS2 and LQ9 blocks feature a 4.000-inch bore and accept heads designed for LS1, LS6, LQ9, L92, LS3, and LS9 engines.

Other considerations involved in selecting the best cylinder head for your application include the weight of the vehicle, camshaft, intake manifold, exhaust system, transmission, gear ratio, drive tire diameter, and intended use. Aftermarket cylinder head manufacturers offer all three port designs, but with a range of intake and exhaust port volumes beyond what the factory heads provide. It's best to discuss your build with the head maker for help in choosing a head. For instance, steeper gearing and/or high camshaft lift will likely require larger port volumes.

Trick Flow's GenX 255 head is designed for six-bolt-per-cylinder block applications. Note the "extra" outboard bolt bosses.

Trick Flow Gen X 255 Head for LS3

PN TFS-3261T003-C01
Fully assembled

- Combustion chamber volume 69 cc
- CNC-machined
- Intake runner volume 255 cc
- Exhaust runner volume 69 cc
- Heart-style combustion chamber
- Square intake port shape
- Standard intake port location
- Oval exhaust port shape, standard location
- Angled spark plug ports
- Intake valve diameter 2.165 inches
- Exhaust valve diameter 1.600 inches
- Maximum valve lift .625 inch
- Valve spring outside diameter 1.300 inches
- Dual valve springs
- Titanium retainers
- 7-degree locks
- Intake valve angle 12 degrees
- Exhaust valve angle 12 degrees
- Ductile iron valve seats
- For 6-bolt-per-cylinder LSX-based engine blocks

The GenX 255 head features rectangular intake ports, CNC-finished. Airflow tests with a bore size of 4.065 inches are impressive. At .400-inch lift, intake flow was rated at 294 cfm and exhaust flow at 215 cfm. At .600-inch lift, intake flow was 363 cfm, with exhaust flow at 252 cfm. At a whopping .700-inch lift, flow rose to 383 and 258 cfm respectively.

Trick Flow

Trick Flow's line of LS heads offers a dizzying array of cylinder head configurations in LS1, LS6, and LS2 cathedral-port and Gen IV rectangular-port designs. Offerings include as-cast and CNC-machined versions, in both bare and fully assembled configurations in a wide range of intake port volumes.

One example is its GenX 255 head for LS3 applications, PN TFS-3261T003-C01.

Dart

Dart offers both LS1-based (cathedral-port intakes) and Gen IV LS3–compatible versions, as well as a dedicated racing head. Its LS1-, LS6-, and LS2-based designs include the Pro 1 15-degree head in three different versions: 205-, 225-, and 250-cc intake runners. The Gen IV Pro 1 LS3–style head offers 280-cc intake runners and a six-bolt-per-cylinder fastener layout. Targeted at serious drag race applications, the Gen IV LS3–compatible Race Series rectangular-port head features 10-degree valve angles and a whopping 280-cc intake runner volume. The competition-only heads are unique, featuring 10-degree canted valves, with intake and exhaust valve locations reversed, and huge oval-port 368-cc intake runners. Dart's line is designed for bore sizes of 4.125 inches or larger and for six-bolt-per-cylinder blocks.

Dart begins its offerings with the Pro 1 205-cc or 225-cc LS1, LS6, and LS2 head, equipped with beehive springs and cathedral intake ports. (Photo Courtesy Dart)

Dart's Gen IV LS3–style head is intended for 4.125-inch bores or larger and accommodates six-bolt-per-cylinder blocks. Valve angle is 10 degrees, with canted valves. Intake and exhaust valve locations are reversed. It features 368-cc intake runners with oval ports. This is a serious race-only drag head. (Photo Courtesy Dart)

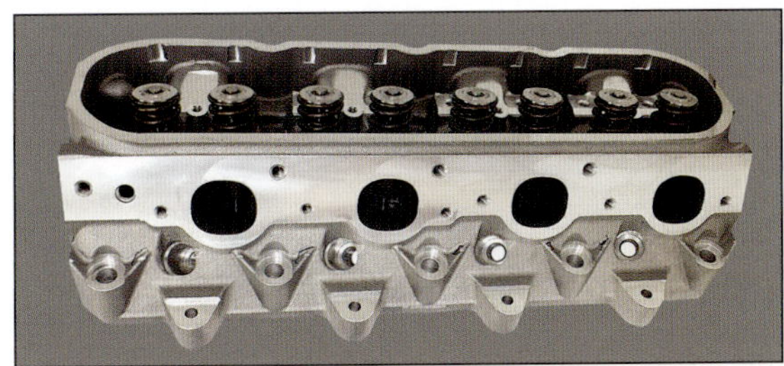

The Pro 1 LS3 head features a six-bolt-per-cylinder format, with 280-cc intake runners and rectangular intake ports. (Photo Courtesy Dart)

Dart Pro 1 225-cc Cylinder Heads

PN 11010010

- Combustion chamber volume 62 cc
- Material 355-T6 aluminum alloy
- Intake port volume 205 cc and 225 cc
- Exhaust port volume 85 cc
- Manifold style LS6
- Intake valve diameter 2.02 inches and 2.05 inches
- Exhaust valve diameter 1.600 inches
- Retainers 7-degree titanium
- Spring pockets 1.350-inch diameter
- Springs 1.290-inch Beehive
- Valve length:
  4.900-inch intake
  4.940-inch exhaust
- Valve stem diameter 8 mm
- Valve guides .439-inch OD manganese bronze
- Valve guide length 2.100 inches
- Valve seats powdered metal .006-inch press fit
- Valve seat dimension
  intake 2.160-inch x .350-inch
  exhaust 1.650-inch x .350-inch

AFR Mongoose LSX 215 Head

- Combustion chamber volume 65 cc
- Intake runner volume 215 cc
- Exhaust runner volume 85 cc
- Stock intake port location
- Stock exhaust port location
- Valve angle 15 degrees
- Valve spring pocket diameter 1.510 inches
- Maximum valve spring pocket machining 1.525 inches
- Deck thickness 0.750 inch
- Minimum bore diameter 3.900 inches
- Pac Racing springs accommodate .600-inch lift
- Valve spring upgrade to handle .650-inch lift is available
- Exhaust valve upgrade available with Inconel valves

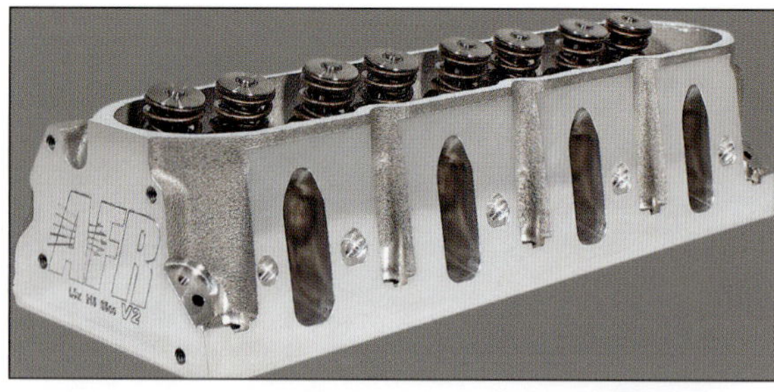

AFR's LSX Mongoose line of cylinder heads offers a selection of cathedral-port LS1 and LS2 heads, including intake runner volumes of 210, 215, 230, and 245 cc. Pictured here is the 215-cc version. (Photo Courtesy AFR)

Air Flow Research (AFR)

AFR's Mongoose line includes both cathedral-port-design LS1 and rectangular-port-design LS3. All heads are 100-percent port and chamber CNC-machined and include titanium retainers. The Gen III LS1 line offers intake runner volumes of 210, 215, 230, and 245 cc. All feature 15-degree valves.

The LS3-style head features 12-degree valves and 260-cc intake runner volume. It has 69-cc combustion chambers with 2.165-inch intake valves and 1.600-inch exhaust valves, dual 1.270-inch valve springs rated at 155 pounds at the seat, 8-mm bronze valve guides, and ductile iron valve seats.

MAST Motorsports

Mast's Black Label line of LS heads offers both cathedral-port and rectangular-port heads. All Black Label heads feature a deck thickness of .750 inch. Mast offers a wide range of configurations.

In the LS1 and LS2 cathedral-port family, the following are available:

- LS1/LS2 225 cc for 3.890+ bore
- LS1/LS2 245 cc for 4.000+ bore
- LS1/LS2 275 cc for 4.125+ bore
- LS1/LS2 295 cc for 4.125+ bore

MAST Motorsports LS3 280-cc Head for 4.125-inch Bore

- Combustion chamber volume 70 cc
- Intake runner volume 280 cc
- Valve angle 12 degrees
- Intake valve diameter 2.20 inches
- Exhaust valve diameter 1.600 inches
- 6-bolt-per-cylinder format
- Deck thickness 0.750 inch
- Valve spring options available for .650- to .750-inch lift
- Accepts LS3 and L76 intake manifolds
- Accepts Gen IV LS7 OEM rockers
- Intake valve options include solid stainless steel, hollow stainless steel, or titanium
- Exhaust valve options include solid stainless steel, Inconel, or titanium

Mast Motorsports' Black Label LS3 head is optimized for LS3 supercharger applications, with a 4.125-inch cylinder bore. Intake runners feature a volume of 280 cc. (Photo Courtesy Mast)

Combustion chambers on the LS3 280-cc head offer 70-cc volumes, with 12-degree valves. (Photo Courtesy Mast)

In the LS3 style range, the following are available:

- LS3 240 cc for 3.890+ bore
- LS3 280 cc for 4.125+ bore
- LS3 LSA 255 cc for 4.000+ bore

In the LS7 style, offerings include the following:

- LS7 265 cc for 4.000+ bore
- LS7 285 cc for 4.125+ bore
- LS7 305 cc for 4.125+ bore

In addition to the cathedral- and rectangular-port versions, Mast also offers an "all-out" race-only cylinder head called the Str8jacket Head, in a medium format for bore sizes of 4.000 inches or larger and the large for bores of 4.125 and larger. Both versions feature 47-cc combustion chambers and inline 11-degree valve layout. The medium bore head is equipped with 2.250-inch intake valves and 1.570-inch exhaust valves. The large bore head uses 2.300-inch intake valves and 1.600-inch exhaust valves. Valve spring options are based on specific lift requirements. Valves are offered in either titanium/titanium or titanium/solid stainless-steel packages.

Edelbrock

Edelbrock offers cathedral port heads for LS1, LS6, and LS2 Gen III and IV applications and a Gen IV LS3–style head. The LS3 head is offered to fit stock LS blocks with a four-bolt-per-cylinder layout or LSX-style blocks with six bolts per cylinder. The LS1, LS6, and LS2 head is offered with an intake runner volume of 215 cc, while the LS3 head features 230 cc.

Edelbrock E-CNC

Edelbrock E-CNC 215 for LS1/LS2

Applies to 4.8L, 5.3L, 5.7L, and 6.0L blocks

- Combustion chamber volume 65 cc
- Intake runner volume 215 cc
- Exhaust runner volume 76 cc
- Intake valve diameter 2.02 inches
- Exhaust valve diameter 1.57 inches
- Valve stem diameter .8 mm
- Deck thickness .625 inch
- Valve spring diameter .55 inch/1.30 inch
- Pushrod diameter .3125 inch
- Valve angle 15 degrees
- Stock exhaust port location

Edelbrock E-CNC 230 for LS3

- Combustion chamber volume 69 cc
- Intake runner volume 230 cc
- Exhaust runner volume 80 cc
- Intake valve diameter 2.135 inches
- Exhaust valve diameter 1.555 inches
- Valve stem diameter .8 mm
- Valve guides in manganese bronze
- Deck thickness .625 inch
- Valve spring diameter 1.30 inches
- Valve spring maximum lift .680 inch
- Pushrod diameter .3125 inch
- Valve angle 13.25 degrees
- Exhaust port location .125-inch raised port

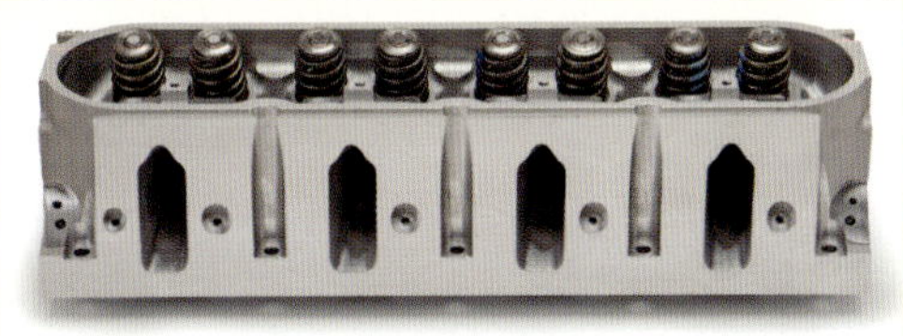

Edelbrock offers two basic head formats for Gen III or Gen IV, including the E-CNC 215-cc head for the cathedral-port LS1/LS6/LS2 version, shown here. (Photo Courtesy Edelbrock)

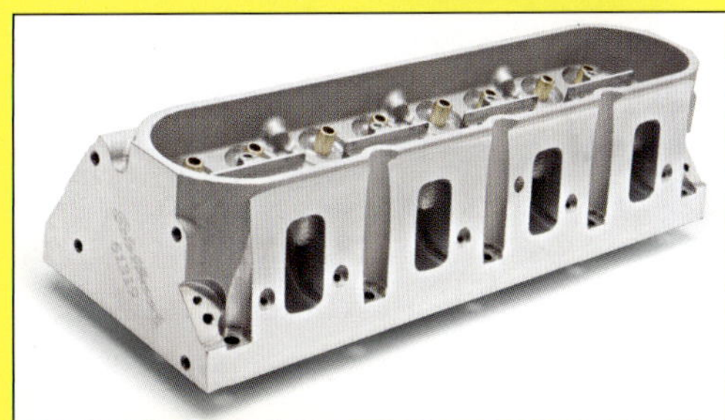

The Edelbrock Gen IV LS3 head features 230-cc intake runners with LS3-style rectangular intake ports. (Photo Courtesy Edelbrock)

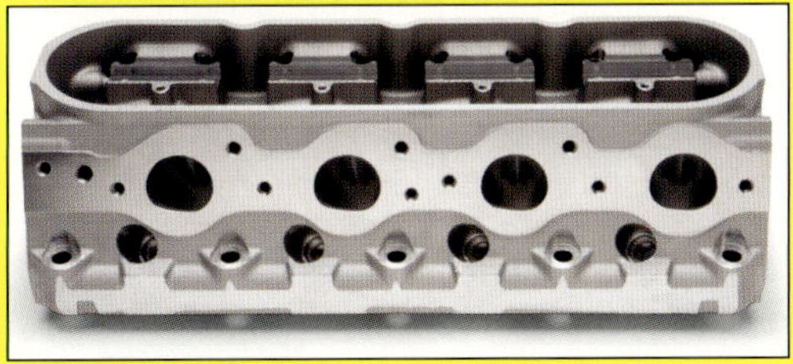

The LS3 head is available for both four-bolt main bearing standard LS blocks and for six-bolt-per-cylinder LSX-style blocks. The four-bolt version is shown here. (Photo Courtesy Edelbrock)

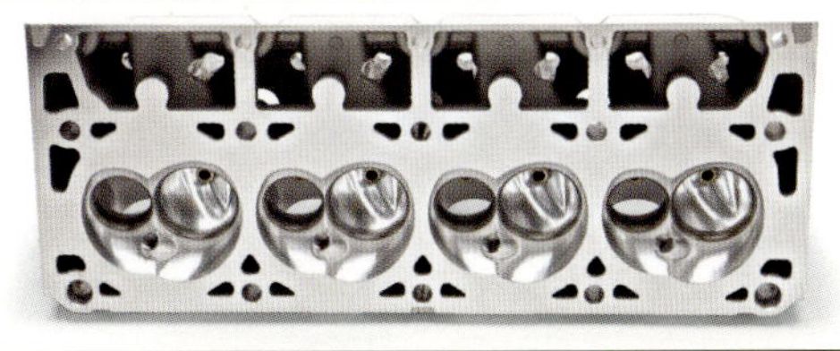

Heart-shaped combustion chambers on the LS3 heads feature a volume of 69 cc. (Photo Courtesy Edelbrock)

Racing Head Service (RHS)

RHS's Pro Elite line of LS heads includes a hefty array of LS7-style versions, with one even applicable to small-bore LS1/LS2 blocks. Rather than featuring the traditional cathedral intake ports found on other LS1/LS2 heads, the small-bore LS7 head features LS7-style rectangular intake ports, providing a power upgrade that allows earlier engines to take advantage of the PS7 head design. This small-bore head features 260-cc intake runners that are raised .220 inch for a straighter shot to the cylinder.

In addition to the small-bore version, seven versions of the LS7 larger-bore heads are available, all with 29-cc intake runner volume. Differences are chamber volumes, types of valves and springs, and applications for hydraulic or solid lifters.

All heads in the Pro Elite series feature a six-bolt pattern, which allows installation in either four-bolt or six-bolt blocks. Heads are available bare or fully assembled.

Bill Mitchell Products (BMP)

Bill Mitchell Products (BMP) offers a killer LS7X head featuring hardened seats; bronze valve guides; spring seats machined for 1.560-inch springs (can be machined for 1.625-inch); Manley stainless-steel valves; 2.250-inch intake valves and 1.625-inch exhaust valves; CNC porting; 12-degree valve angle; flat-machined rocker pedestals that accept T&D rocker system 2351; 285-cc or 296-cc rectangular-port intake runners; 106-cc or 112-cc exhaust runners; 64-cc combustion chambers as cast, or 74-cc CNC; and a valve cover rail raised .300 inch for added rocker clearance. The head features a six-bolt head bolt pattern, allowing installation to four- or six-bolt blocks.

RHS LS7 Pro Elite 291

- Combustion chamber volume 69 cc
- CNC-ported
- Titanium and stainless-steel valves
- Titanium retainers
- Intake valve diameter 2.200 inches
- Exhaust valve diameter 1.615 inches
- Exhaust ports .100-inch raised work with stock manifolds
- Maximum lift .675 inch
- Valve angle 12 degrees
- Deck thickness 0.750 inch
- 0.400-inch raised and reinforced rocker rail works with aftermarket rockers

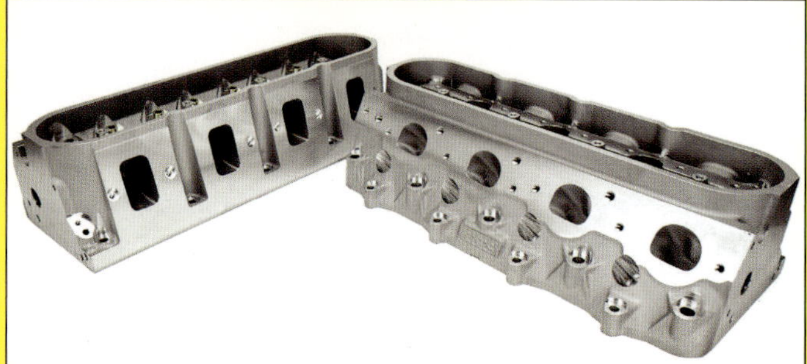

RHS's Pro Elite heads feature LS7-style rectangular-intake ports and six-bolts-per-cylinder bolt holes, even in its small-bore head intended for LS1/LS2 applications. (Photo Courtesy Comp Cams)

BMP LS7x Cylinder Heads

PN 025350C-0

- Bare (fully assembled available)
- Combustion chamber volume 64 cc
- CNC-machined
- Intake runner volume 285 cc or 296 cc
- Exhaust ports 106 cc or 112 cc
- Material 355-T6 alloy high-density aluminum
- Intake and exhaust hardened valve seats
- Valve guides manganese bronze
- Intake valve diameter 2.250-inch
- Exhaust valve diameter 1.625-inch
- Valve angle 12 degrees
- Valve seat angle 50 degrees
- Spring seats machined for 1.560 inches (can be machined to 1.650 inches)
- Valves (in assembled heads) Manley stainless steel
- Rocker arms (in assembled heads) Jesel KPS499187 or T&D 2351
- Spark plug 14 mm, .708-inch reach, taper seat
- Raised valve cover rails

BMP's LS7X head is offered with either 285-cc or 296-cc intake runner volume, and with either as-cast 64-cc combustion chambers or CNC'd 74-cc chambers. The rocker rail pedestals feature radiused pockets to accept a T&D rocker system. (Photo Courtesy BMP)

Pistons

The OEM pistons for the LS are all cast hypereutectic construction except for the LS9 with its factory-forged pistons. Don't be alarmed about the "cast" term. Unlike earlier cast pistons, hypereutectic pistons feature a dense casting process with high silicon content. They're stronger than old-school cast pistons, and they're more stable in terms of thermal expansion and contraction. The dimensional stability allows the factory to run tighter bore clearance. Hyper pistons are fine for up to about 500 to 550 hp generally. Beyond that, or if combustion pressures will rise due to the use of forced induction (supercharging or turbocharging) and/or nitrous oxide injection, you'll need to upgrade to forged pistons.

Regardless of the application, if you're building or rebuilding an LS engine from scratch and intend to boost horsepower, it just makes sense to purchase a set of forged pistons to eliminate the potential variable. Aftermarket performance piston manufacturers offer a wide selection of forged pistons for the LS engine format, in a variety of popular bore sizes, compression heights, and compression ratios. Check brands such as JE, Wiseco, Diamond, Ross, CP, and others.

Piston Compression Height

Compression height, also called compression distance or piston CD, refers to the wrist pin bore centerline to the piston deck. If you're planning to build a stroker engine, the combination of crankshaft stroke, block deck height, connecting rod length, and piston compression height is relative to the piston deck's location to the block deck at top dead center (TDC).

As an example, let's say that the block deck height is 9.240 inches, which is the GM factory spec for LS deck height. Block deck height refers to the distance from the main bore centerline to the block deck. However, OEM blocks are rarely machined properly. The raw block may be slightly less or slightly more than 9.240 inches, decks may be finished nonparallel to the main bore centerline, and deck height may differ from the front to the rear of the block. The block should be corrected prior to selecting components to verify the final corrected deck height. But for the purposes of theory, let's say that our block's deck height is 9.240 inches.

For the purposes of this example, our crankshaft stroke is 4.000 inches. To determine where our piston dome will be located relative to the block deck, we need only factor half of the crankshaft stroke, from the rod journal's top-dead-center location to the block deck. So, in this case our stroke factor is 2.000 inches. At this point we know that 2.000 inches of our 9.240-inch deck height is taken up by the crank stroke. The remaining theoretical distance of 7.240 inches must be achieved by the combination of our connecting rod length and our piston compression distance. Rod length refers to the distance from the centerline of the rod big end to the centerline of the rod's small-end bore. By selecting a rod length of 6.125 inches, this leaves a required piston compression distance of 1.115 inches. Formula:

1/2 Stroke + Rod Length + Piston CD
= Block Deck Height

In this example:

2.000 + 6.125 + 1.115 = 9.240

This would theoretically place our piston dome flat flush with the block deck.

With block deck height, crank stroke, and rod length already known, to determine piston CD is easy using the following formula:

Block Deck Height – (1/2 Stroke + Rod Length) = Piston CD

Using the previous example:

9.240 – (2.000 + 6.125) = 1.115

Aftermarket performance forged pistons may be ordered in a range of compression distances. Some piston makers offer a specific range of compression heights, while some makers will custom-machine your pistons to whatever compression height you require, as long as the pin bore will not intersect the second compression ring groove.

If the block decks have been machined to a height less than 9.240 inches, the pistons may protrude a few thousandths above the decks. To obtain adequate piston-to-valve clearance, these clearances must be checked, factoring in total valve lift and cam duration, rocker arm ratio, intake and valve diameters for radial clearance, and thickness of the cylinder head gasket. If during test fitting and measuring, valve clearance is not adequate, a thicker head gasket may be selected to compensate. Another option is to have the piston dome's valve pockets milled to a larger radius if the valve head diameter is a tick too large for radial clearance. As far as valve-to-piston clearances are concerned, a minimum recommended clearance for intake valves should be .080 inch, with a minimum of .100 inch for the exhaust valves. This clearance is even more of a concern with an aluminum block, which has a greater potential for thermal expansion.

When checking valve clearance with head gaskets installed, the gaskets must be crushed as they would be during final assembly. Multi-layer steel (MLS) head gaskets can easily be measured for thickness by measuring thickness adjacent to the gasket rivets, where the layers are already fully compressed. MLS gaskets are available in a variety of thicknesses, generally in the .041- to .045-inch range. However, some gasket makers, such as Cometic, can supply custom MLS gaskets in a much wider range of thicknesses. For example, if a compressed gasket thickness of .045 inch does not provide adequate clearance, a gasket thickness of .051 inch may be obtained. This is just an example.

Depending on stroke and rod length, a shorter piston CD may be required to keep the piston flush or just below the block deck. In many cases, shortening the piston CD,

Before ordering oversize pistons, it's wise to measure the cylinder wall thickness in all cylinders using an ultrasonic thickness gauge. This will aid in determining how far the boring/honing oversize can go. Measure at a variety of bore height and clock positions, as wall thickness may vary.

To place the piston at TDC relative to the block's deck height, with certain crankshaft stroke and rod length variables, the piston compression height may dictate that the pin bore be raised, intersecting the oil ring groove. This requires the use of an oil ring support rail that completes the footprint for the oil ring package over each end of the pin bore.

A dial indicator aids in measuring the precise top-dead-center (TDC) position of the piston. With the piston at TDC, a depth micrometer can be used to measure deck height relative to the block deck, to determine if the piston is below, flush with, or above deck.

The support rail features a male pimple that prevents the rail from rotating and may prevent placing the rail gap at one of the two relief areas. When installing the rail, the male dot must be placed at either relief opening. Support rails are included in the oil ring package for pistons that require these rails.

which involves moving the wrist pin bore upward, results in the pin bore intersecting with the oil ring groove. In this case, the oil ring groove will have been machined taller to accommodate both the oil ring package and a "support rail," which serves to complete the footprint for the oil ring package at each side of the wrist pin bore.

An example of a flattop piston dome. Depending on the cylinder head combustion chamber volume, choosing piston dome volume in combination allows you to tailor the desired compression ratio.

High-dome pistons obviously create a smaller combustion chamber volume, increasing compression ratio compared to a flattop dome.

Bore Size, Dome, and Compression Ratio

The available combinations of cylinder bore diameter, piston dome shape, and static compression ratio for LS engine applications are quite extensive. The standard bore size for LS1 and LS6 formats is 3.898 inches. The standard bore size for LS2 is 4.000 inches. Bore size for LS3 is 4.065 inches, and LS7 standard bore size is 4.125 inches. Oversizes are available in a variety of diameters, depending on whether the block is a standard bore and whether it is an iron or an aluminum block. While iron blocks allow a degree of overboring, aluminum blocks with their integrally cast-in sleeves allow for only a slight bit of honing oversize. For example, an aluminum LS1 or LS6 block will accept an oversize up to only around 3.905 inches. Aluminum LS2 blocks will accept honing oversize to about 4.030 inches. LS3 aluminum block cylinder bore oversizing is limited to about 4.080 inches. LS7 blocks may be oversized to about 4.130 inches. Iron blocks may accept a further increase, but cylinder walls should first be measured for thickness using an ultrasonic thickness gauge to verify the minimum available wall thickness. Finished wall thickness generally should be limited to a minimum of about .180 to .200 inch. Again, as a generalization, an LQ9 6.0L iron block should accept an overbore of .060 inch, although an overbore of .030 inch would be safer.

If you plan to use forced induction to boost cylinder pressure, stay away from a theoretically minimum wall thickness. The thicker, the better. Because of the tolerances involved in both aluminum block cast-in cylinder liners and iron block castings, cylinder wall thickness may not be uniform from top to bottom or around the circumference of the bores. Before oversizing, measure existing wall thickness from top to bottom and at a variety of clock positions on each cylinder location.

Off-the-shelf dome configurations include flattop, inverted dome, and domed. Flattop pistons will feature intake and exhaust valve relief pockets. Inverted dome pistons feature a relieved dome "bowl," while domed pistons feature a raised dome area. Obviously, the volume of the dome affects compression ratio. Inverted domes provide a lower compression and domed provide a higher compression ratio. Obtaining the desired compression ratio is not limited to the piston itself. The combination of piston dome volume and cylinder head combustion chamber volume work in unison to create static compression ratio.

Piston makers offer comprehensive charts that list the cubic-inch displacement and static compression ratio for any given combination of cylinder bore diameter, crank stroke, rod length, piston compression height, block deck height, piston dome volume, and combustion chamber volume.

While we don't have the space here to list all possible combinations for all LS platforms, citing an LS2 aluminum block as an example, a cylinder bore oversize of 4.005 inches coupled with a crank stroke of 4.000 inches would provide 403 ci of displacement. To mate with the 9.240-inch block deck height, rod length would be 6.125 inches and piston compression height would be 1.115 inches.

Using a flattop piston with a dome volume of –5 cc and a cylinder head with a combustion chamber volume of 64 cc would provide a compression ratio of approximately 11.6:1. With a combustion chamber volume of 68 cc, compression would be about 11.3:1. With a 70-cc

An example of an inverted dome piston for an LS3 application. The JE forged piston shown here features a .097-inch-deep inverted dome and 12-degree valve relief pockets. The lines you see here are simply surface shadows left by the CNC-milling under studio lighting. The dome surface is actually flat and smooth. Dome volume is –14.6 cc. Using an inverted dome allows a reduction of compression ratio compared to a flattop piston.

combustion chamber, compression would be about 10.8:1.

Using the same combination of parts but with an inverted dome piston that has –7.6 cc of volume, using a 64-cc combustion chamber would provide about 10.9:1. With a combustion chamber of 68 cc, compression would be about 10.5:1. With a 70-cc combustion chamber, compression would be about 10.3:1.

Again, using the same above combination but with a domed piston that has 5.0 cc of volume, a 64-cc chamber would provide about 13.2:1. A 68-cc chamber would provide about 12.5:1, and a 70-cc chamber would provide about 12.2:1. Depending on bore size, piston dome configuration, and cylinder head combustion chamber volume, pistons for LS applications may be selected to achieve anywhere from about 8.1:1 to 13.2:1 static compression ratio.

Keep in mind that if you plan to run forced induction, the static build compression ratio will dramatically increase at effective compression when boost is added. For example, with a Roots-style supercharger, a static ratio of 8.0:1 can rise to an effective ratio of 11.3:1 with only 6 psi of boost, and to more than 15.6:1 with 14 psi boost. Generally speaking, for street applications running aluminum heads and 92 octane fuel, you should try to keep effective compression to about 12:1 maximum. Suggested limit examples are as follows:

Piston Selection of KS7 Flattop Pistons

To demonstrate piston selection based on stroke, bore diameter, rod length, piston compression height, and cylinder head combustion chamber volume, following are examples of LS7 flattop pistons that feature a dome volume of –2 cc. Note: "CD" stands for piston compression distance and "CR" stands for compression ratio. All results are based on a block deck height of 9.240 inches. To illustrate one piston maker's offerings, all examples here are listings by JE Pistons.

LS7 Flattop

Piston ci	Bore	Stroke	Rod Length	CD	CR with 70-cc Head
438	4.125	4.100	6.125	1.065	12.0
439	4.130	4.100	6.125	1.065	12.0
427	4.125	4.000	6.077	1.163	11.7
428	4.1275	4.000	6.125	1.115	11.7
428	4.1275	4.000	6.077	1.163	11.7
429	4.130	4.000	6.077	1.163	11.7
429	4.130	4.000	6.125	1.115	11.7

The following examples are based on LS2/LS3 builds, with choices of three different cylinder head combustion chamber volumes. Bear in mind that an incredibly diverse selection of pistons is available for these applications. Here I've listed only a few. These examples feature flattop pistons with a dome volume of –5 cc.

LS2/LS3 Flattop

Piston ci	Bore	Stroke	Rod Length	CD	CR 64 cc	CR 68 cc	CR 70 cc
413	4.000	4.100	6.125	1.065	11.8	11.3	11.1
403	4.005	4.000	6.125	1.115	11.6	11.1	10.8
408	4.030	4.000	6.125	1.115	11.7	11.2	11.0
392	4.000	3.900	6.125	1.165	11.0	10.5	10.3
406	4.070	3.900	6.125	1.165	11.3	10.8	10.6
418	4.130	3.900	6.125	1.165	11.6	11.1	10.8

As noted, the examples from JE Pistons represent only a small number of build variations. Depending on stroke, bore diameter, rod length, piston compression height, combustion chamber volume, and piston dome volume, off-the-shelf forged pistons are readily available to achieve compression ratios of 8.0:1 to 13.2:1. Obviously, flattop or inverted dome pistons and/or larger combustion chambers can achieve lower compression ratios. Whole higher ratios can be had with small combustion chambers and/or higher-domed pistons. ■

Static Compression	Boost	Effective Compression
8.0:1	8 psi	12.4:1
9.0:1	4 psi	12.1:1
9.5:1	4 psi	12.1:1
10.0:1	2 psi	11.9:1
11.0:1	2 psi	12.5:1

Skirt-to-Wall Clearance

The engine block's cylinder bores should never be final-finished until you have your intended pistons in hand. The piston skirt diameter dictates the required finish-honed bore diameter. Measure skirt diameter and add the wall clearance recommended by the piston maker to finish-hone the bores. It is critical to measure the piston skirts at exactly the location specified by the piston maker because piston skirts feature a very slight taper. As an example, JE forged pistons are usually measured exactly .500 inch from the bottom of the skirt, unless the measuring location is specified otherwise. The measuring point varies depending on the specific piston design. Add to this the recommended piston clearance to final-hone the cylinders. It's important to note that piston diameters are not measured at the ring area, but at a specific point on the skirts. A general rule of thumb for forged pistons is .001-inch clearance for every inch of cylinder bore diameter. For example, a 4.000-inch bore would require .004-inch piston-to-wall clearance. Again, this is a generic specification. *Always* adhere to the clearance recommendation provided by the piston maker.

Before finishing the cylinder bores to size, the piston skirt diameter must be measured to determine the bore diameter required for the recommended piston-skirt-to-wall clearance. It is critical to measure only at the skirt area specified by the piston maker.

Piston and Rod Orientation

The installed direction of pistons to connecting rods is critical. Let's consider the rod orientation. If the big end of the connecting rod features a larger chamfer on one side, this side must be installed facing the crankshaft's journal radius fillet. If the rods are designed for use on a crank that does not feature a radiused fillet, the rods may not feature a large chamfer on one side.

If there is no noticeable chamfer on either side of the rod big end, the bearing placement on the side of the rod that faces the fillet should be slightly spaced away from the fillet to prevent the bearing from digging into the fillet radius.

Valve pockets can be used as reference as well when orienting the pistons. The larger intake valve pocket faces the front of the engine on the left-side cylinder head, while the intake valve pocket faces the rear of the right-side head.

Skirts and Major/Minor Thrust

The shape, area of mass, and weight of a piston's skirts play a major role in managing friction and in stabilizing the piston during TDC and BDC transitions. Here we'll discuss the role of the major and minor thrust sides of a piston and the development of asymmetric skirt designs intended to minimize weight while maximizing efficiency.

Piston skirts are not perfectly round, and each side of the piston experiences different levels of loading relative to the intake and exhaust sides of the cylinders. Skirt design plays a major role in accommodating these forces in ways that improve durability and performance; so does piston weight. The piston skirt area is slightly barrel shaped to provide an adequate surface load against the cylinder wall while reducing friction. The amount of surface area must accommodate the load while providing piston stability to minimize

Original equipment hypereutectic flattop pistons feature a dot on the dome. The dot indicates that the piston is installed with the dot facing the front of the engine on the right bank and with the dot facing the rear of the engine on the left bank.

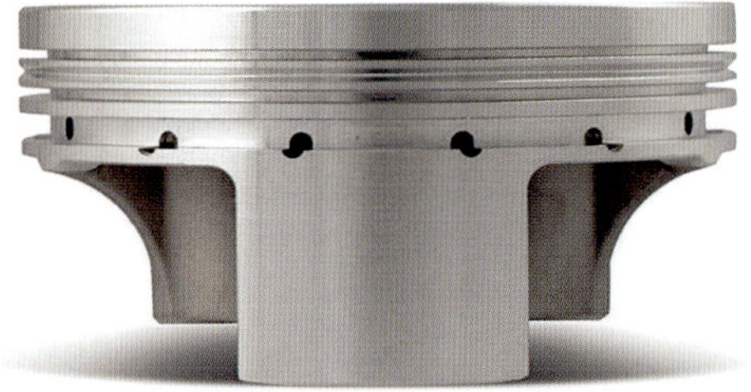

This piston's minor thrust side skirt design provides great contact and support for the thrust side where it's needed, and a small skirt for the minor side, where it's less critical, saving weight in the process.

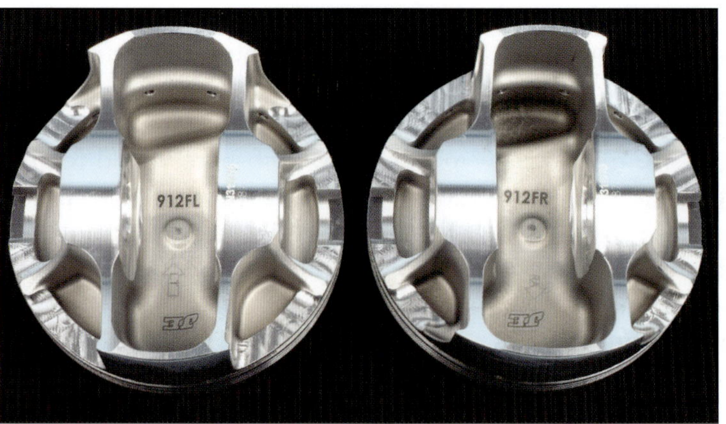

Asymmetric pistons feature a wider skirt at the major thrust side and a smaller skirt at the minor thrust side. The underside view clearly shows the difference.

rocking relative to the pin axis as the piston moves down from TDC and back up from BDC. The piston experiences a "major" and "minor" thrust force at opposing sides of the piston skirts. The major thrust face is the side of the piston that receives the thrust on the power stroke. As viewed facing the front of the engine, if the crankshaft is rotating clockwise, the major thrust face is on the left side of the cylinder (the exhaust sides of the right/passenger-side cylinders; and the intake sides of the left/driver–side cylinders). The minor thrust side experiences force on the compression stroke.

This difference in force at each side of the piston is caused in part by the operating angles of the connecting rod during its travel. During the firing cycle, the load experienced on the major thrust side skirt can be as much as 10 times greater than the load experienced on the minor thrust side skirt. The difference in skirt loading will vary depending on variables such as crankshaft stroke, connecting rod length, and peak cylinder pressures.

Asymmetric pistons are bank specific, and thus each piston is labeled for right or left bank position. The dome may also feature a laser-etched arrow that indicates piston orientation toward the front of the engine.

Major Thrust Side

When the piston is pushed down during the power stroke, it experiences resistance as it attempts to turn the crankshaft. As load increases, the amount of resistance increases. During this resistance, the piston side load is forced to one side, and that's the major thrust side. It places more force and subsequently increased friction and potential wear on the thrust side of the cylinder wall. If the piston dome features a reference dot or other orientation mark, it's critical to install the piston with this mark facing the appropriate direction. Usually the mark indicates the side of the piston that should face forward. The piston side loads on the major side tend to increase with the use of a longer stroke and with forced/boosted induction pressures. Again, assuming a clockwise-rotating crankshaft, the major thrust side will be at the exhaust side of the engine's right bank and the intake side of the left bank.

Minor Thrust Side

The piston's minor thrust side is directly opposite the major thrust side. The minor thrust side is forced to the opposite side of the cylinder wall as it moves up on the compression stroke by the resistance generated by meeting the air/fuel mixture. The role of the minor thrust side is basically to provide piston stability, with the major thrust side taking the brunt of the cylinder wall contact. Due to its "less force" role, the minor thrust side skirt can be narrower, saving weight without sacrificing strength.

To address, or "fine-tune," these forces between the major and minor thrust sides, asymmetric pistons have been developed that feature two different-size skirts.

This style of piston is specifically designed with a larger (wider) skirt on the major thrust side and a smaller skirt on the minor thrust side. This provides a greater "footprint" for the major thrust side, where it's needed the most to handle a higher degree of thrust loading, and allows the piston weight to be slightly reduced by featuring a small footprint on the opposite/minor thrust side, where the force is less. During the power stroke, when the piston changes direction at top dead center, combustion pressure pushes the piston down and at the same time pushes the thrust side of the skirt toward the cylinder wall.

Citing JE Pistons' "asymmetric" design as an example, in its forged side relief (FSR) line, a wider skirt area is featured on the major thrust side, and the pin bosses are relieved at the outboard sides to allow the use of a shorter (and lighter) wrist pin.

The asymmetric design approach was initially developed for specific racing applications, but the concept has trickled down to street applications, with the LS platform as a good example.

Another benefit to the asymmetric approach is increased piston ring sealing and ring stability thanks to the skirt mass and profile. Basically, the dedicated major and minor thrust skirt design coupled with a slightly offset wrist pin directly addresses ring performance in addition to reduced wall friction.

Offset Pin

Asymmetric pistons also feature an offset wrist pin, with the pin centerline biased from zero toward the major thrust side by .020 inch. This slight offset tends to balance the piston to accommodate the difference in skirt mass and to compensate

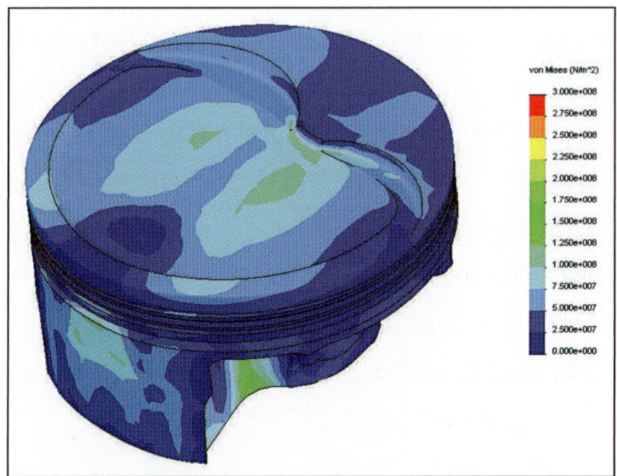

This finite element analysis (FEA) view shows even stress forces at both major and minor thrust sides (note the dark dome areas), even though skirts differ in area. The offset location of the wrist pin aids in balancing out the pivot point. (Photo Courtesy JE Pistons)

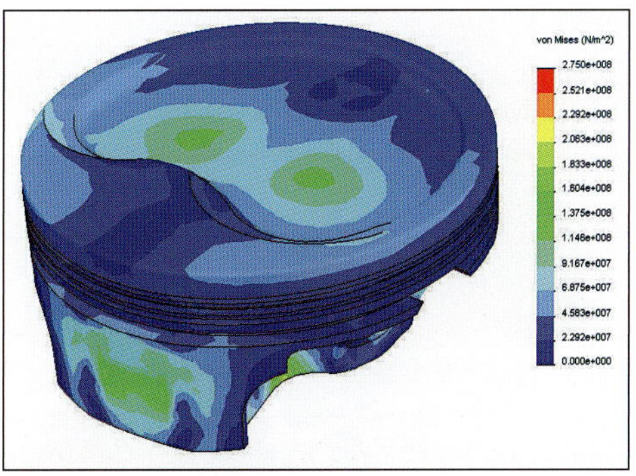

FEA takes a snapshot of the piston's stress levels at the worst-case scenario, which differs greatly from an engine that's at part throttle. The image plots the stress level. The high-stress areas are shown in red in accordance with the chart on the right. This is a simulation of stress under firing. (Photo Courtesy JE Pistons)

for and alter the effect of rod angle, transferring a bit of force away from the major thrust side.

Again citing JE's development in this area, the asymmetric design allows the use of shorter, stiffer, and lighter wrist pins. According to JE, a typical weight savings is about 10 grams.

Note: The contact pressure FEA images here show contact pressure specifically between the skirt panel and the bore. It's important to analyze both skirt profiles on an asymmetrical piston design even though the minor thrust experiences much less pressure. On a symmetrical design, typically only the major thrust is analyzed. Stress images show how the stress at the skirts affects the rest of the piston.

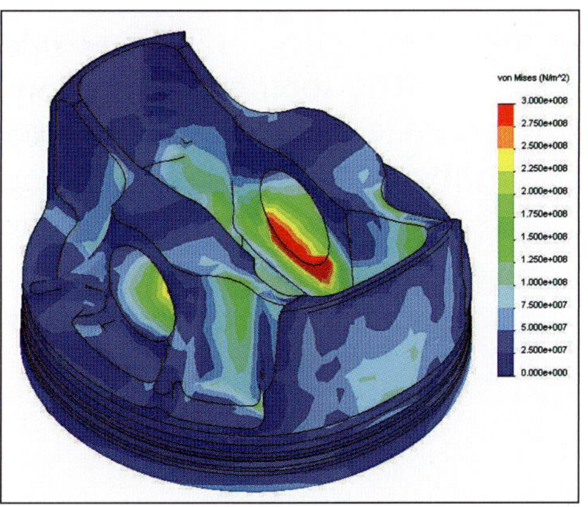

The stress of the piston under major stress. (Photo Courtesy JE Pistons)

Piston Coatings

A variety of coatings is available for both internal and external engine components, with applications varying from power enhancement, engine efficiency, and durability/longevity to corrosion protection and external appearance. Here we'll focus on coatings that apply to pistons. Today's coatings include those designed to reduce friction and enhance lubricity and thermal barrier coatings for increased engine efficiency and protection of components exposed to extreme heat levels.

Thermal Barrier Coatings

Thermal barrier coatings (which are intended to serve as a heat shield) feature a ceramic formulation designed to prevent excess heat from passing into and being absorbed into the piston domes. This theoretically increases combustion efficiency and reduces piston dimensional changes that may occur due to thermal expansion.

The specific formulas differ depending on the coating manufacturer and the application. Specialized thermal barrier coatings applied to piston tops aid in reflecting heat, reducing the amount of piston expansion (allowing the builder to maintain more consistent wall clearance), and protecting the piston from extreme temperatures encountered in forced-induction systems. For forced-induction and nitrous applications, highly specialized thermal barrier coatings are available not only to enhance combustion efficiency but also to protect the piston from potential heat-related damage.

Bear in mind that the quality of application of a ceramic coating (especially for internal engine components) is extremely critical. The last thing you want is for hard and potentially damaging ceramic to break loose and contaminate the engine. That's why you need to use established coating services, such as those offered by Swain Tech Coatings, Polydyn, and others. When properly applied, the coating not only adheres to the applied surfaces but permanently bonds to the material, essentially becoming part of the base material. If substandard application practices are used, the coating may not be fully bonded and could flake off. In other words, don't try this at home.

While building engines for 24-hour endurance racing, I've had many dozens of engine build components ceramic coated (heads, pistons, valves, exhaust) with absolute success, but I've seen a few devastating issues with components that were "coated" by inexperienced mom 'n' pop shops that caused serious damage. Don't be afraid of ceramic internal coatings; the top shops do an outstanding job, and they know what they're doing. Don't be tempted to go for bargain-basement services.

Antifriction piston skirt coatings are available already installed on new pistons, as shown in this example, or by sending your pistons to a specialty coating service.

Specialty coating options for pistons include a thermal barrier dome coating, antifriction skirt coating, or both.

AntiFriction Coatings

Antifriction coatings are applicable to surfaces that make or potentially make contact, such as (but not limited to) main bearings, rod bearing, cam bearings, and piston skirts. Antifriction coatings, also called lubricity coatings, provide a temporary lubricity when/if the surface is starved for oil, upon cold starts, and during piston rock when transitioning from top dead center to bottom dead center. This type of protective coating also serves to improve oil retention on the surface. While specific antifriction coating formulas vary among the coating services, the materials are generally composed of moly-, graphite-, or Teflon-based materials. The intent is to provide better oil retention and to provide a super-slippery surface.

Piston skirt coatings are generally applied at an average thickness of about .0005 inch per side, which might provide about a .001-inch increase in piston overall skirt diameter. The moly has been applied in such a thin layer, no additional bore dimension changes are required to run "moly-coated" pistons. Unless otherwise instructed by the piston maker, *do not* compensate for the added moly coating when finishing your bores. A specialty coating service may apply antifriction piston skirt coatings or, depending on the piston manufacturer, new pistons are often available with the coatings already applied. While you may or may not need a skirt coating, there's no downside to this application. A skirt coating won't hurt, and it may very well help to prevent skirt wear when called upon.

Camshafts and Lifters

Today's performance aftermarket offers a wide selection of camshaft profiles for enhanced performance, from street-mild to all-out racing competition. While factory camshafts all utilize hydraulic roller lifters, the performance aftermarket offers lifter application in both hydraulic and solid designs. One very significant feature of the LS cam design is its cylinder firing order. While the Gen I small-block Chevy cam firing order is 1-8-4-3-6-5-7-2, the LS firing order was changed to 1-8-7-2-6-5-4-3 to enhance power. This is discussed later in this chapter, as are the theory and function of factory variable valve timing and displacement on demand, two different approaches that GM employed to improve fuel economy.

Camshaft Terminology

A brief explanation of camshaft terminology and specifications follows here, citing information provided by Lunati as the examples.

Nose

The camshaft lobe nose is the highest point that provides maximum valve lift, between the opening and closing ramps of the lobe.

Base Circle

The base circle, also referred to as the lobe heel, is the lowest point of the lobe. This is the point at which the valve is in the fully closed position and the point at which valve lash adjustments are made. When measuring for a custom pushrod length, the lifter is at its lowest point, where pushrod length is measured between the lift cup and rocker arm.

Symmetrical Lobes

Lobes that are symmetric are machined with mirror-image/identical opening and closing ramps on each side of the lobe.

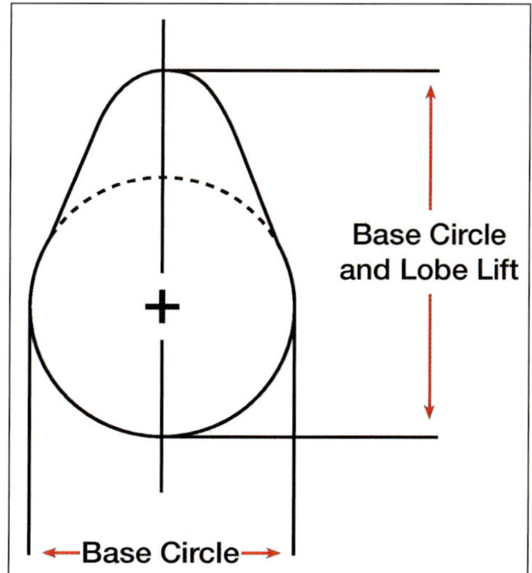

Asymmetrical Lobes

Asymmetrical lobes are machined differently at the opening versus closing ramps. Typically, an asymmetric lobe provides a faster/higher-velocity opening ramp and a slower-velocity closing ramp. This allows the valve to close with less impact force.

Lift

Lift refers to how far the valve is raised from its seat when completely open. Note that lobe lift and valve lift differ. Lobe lift refers to the distance from the base circle to the lobe peak. Valve lift considers both lobe lift and rocker arm ratio. Valve lift is easily calculated by simply multiplying the lobe lift by the rocker arm ratio. For example, if the cam features a lobe lift of .377 inch and is coupled with a rocker arm ratio of 1.7:1, the effective valve lift is .641 inch. If a 1.8:1 rocker arm ratio is used with the same cam, the effective valve lift is 0.678 inch.

The intake and exhaust valves

The camshaft base circle, or heel, is the lowest point of the lobe, opposite the lobe peak. When measuring for pushrod length, the lifter must be on the base, at which point the valve is fully closed. (Photo Courtesy Lunati)

need to open to let air/fuel in and exhaust out of the cylinders. Opening the valves quicker and farther usually will increase engine output. Increasing valve lift without increasing duration can yield more power without much change to the nature of the power curve. However, an increase in valve lift almost always is accompanied by an increase in duration because the lobe ramps are limited in their shape, which is directly related to the type of lifters being used, such as flat tappet or roller.

While many cams may feature more lift on the exhaust side than the intake side, this is primarily done to help compensate for less-than-efficient exhaust ports. Today's high-performance LS heads feature improved flow, so you'll see cams available with more intake lift to help draw more mixture into the chambers.

Duration

Duration represents the angle in crankshaft degrees that the valve stays off its seat during the lifting cycle of the camshaft lobe. Increasing duration keeps the valve open longer and can increase high-RPM power. This also increases the RPM range where the engine produces power. Increasing duration without a change in the lobe separation angle (LSA) will result in increased valve overlap.

When you view cam specifications, advertised duration and duration at .050 inch will be listed. This is the angle in crankshaft degrees that the lifter is lifted more than a predetermined amount off its seat. Because advertised duration can vary depending on cam makers, based on a predetermined point chosen by the cam maker, an industry standard of listing duration at .050 inch allows direct comparison of duration across all brands. Duration at .050 is a measurement of the movement of the lifter, in crankshaft degrees, from the point where it first lifts .050 inch from the base circle on the opening ramp side of the cam lobe to the point where it ends up being .050 inch from the base circle on the closing ramp's side of the lobe. This is the industry standard and is a good value to use to compare camshafts from different manufacturers.

Lobe Separation

Lobe separation, referred to as LSA, is the angle in camshaft degrees between the maximum lift points of the intake and exhaust valves. Lobe separation affects valve overlap, which affects the nature of the power curve, idle quality, and idle vacuum. LSA can be measured using a dial indicator and a degree wheel, but it is usually calculated by dividing the sum of the intake centerline and the exhaust centerline by two. In very general terms, a low LSA, let's say from 106 to 108 degrees, for example, may produce a rougher idle and lower engine vacuum at idle and move the power band to the mid-to-high RPM range. A larger LSA, let's say in the 112- to 114-degree range, would produce a smoother idle and more engine vacuum at idle and would move the torque curve to the lower RPM range for quicker response and a broader RPM band.

Overlap

Overlap refers to the angle in crankshaft degrees that both the intake and exhaust valves are open. This occurs at the end of the exhaust stroke and the beginning of the intake stroke. Increasing lift or duration and/or decreasing lobe separation increases overlap. At high engine speeds, valve overlap allows the rush of exhaust gasses out of the exhaust valve to help pull the fresh air/fuel mixture into the cylinder through the intake valve. Increased engine speed enhances this effect. Increasing overlap increases top-end power and reduces low-speed power and idle quality. Overlap can be calculated by adding the exhaust closing and the intake opening points. For example, a camshaft with an exhaust closing point at 4 degrees after top dead center (ATDC) and an intake opening point at 8 degrees before top dead center (BTDC) has 12 degrees of overlap. Decreasing the lobe separation by only a few degrees can have a huge effect on the overlap area.

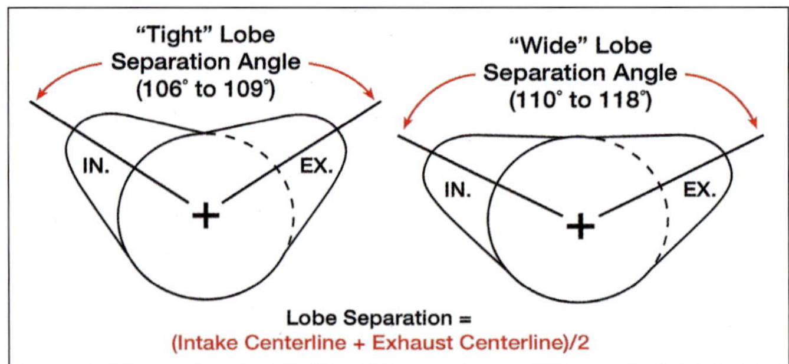

Camshaft lobe separation, referred to as LSA, is the angle in camshaft degrees between the maximum lift points of the intake and exhaust valves. Lobe separation affects valve overlap, which affects the nature of the power curve, idle quality, and idle vacuum. LSA can be measured using a dial indicator and a degree wheel, but it is usually calculated by dividing the sum of the intake centerline and the exhaust centerline by two. (Photo Courtesy Lunati)

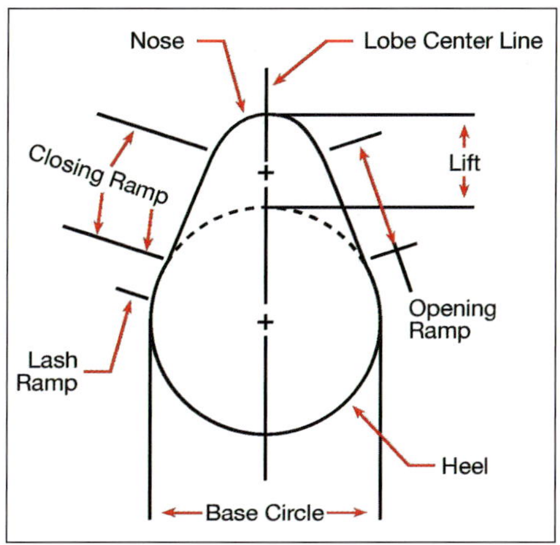

The camshaft centerline is the point halfway between the intake and exhaust centerlines. The intake centerline is the highest point of lift on the intake lobe, expressed in crankshaft degrees after top dead center (ATDC). The exhaust centerline is the highest lift point on the exhaust lobe, expressed in crankshaft degrees before top dead center (BTDC). (Photo Courtesy Lunati)

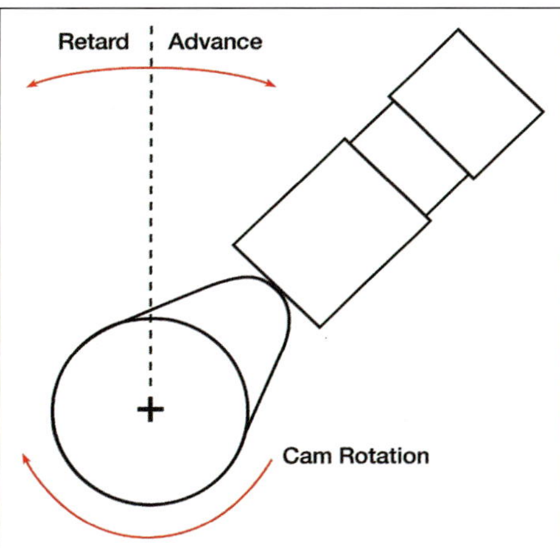

Advancing or retarding the camshaft moves the engine's torque band around the RPM scale by moving the valve events farther ahead or behind the movement of the piston. With the use of an adjustable aftermarket cam timing gear set, you're able to advance the cam to improve low-end power and response, or retard cam timing in favor of high-end power.

Centerline

The intake centerline is the highest point of lift on the intake lobe, expressed in crankshaft degrees after top dead center (ATDC). The exhaust centerline is the highest lift point on the exhaust lobe, expressed in crankshaft degrees before top dead center (BTDC). The camshaft centerline is the point halfway between the intake and exhaust centerlines.

Cam Advance and Retard

Advancing or retarding the camshaft moves the engine's torque band around the RPM scale by moving the valve events farther ahead of or behind the movement of the piston. Typically, a racer will experiment with advancing or retarding a camshaft from the "straight up" location. In general terms, advancing a cam will improve low-end power and response, while retarding a cam biases the torque band to favor high-end power.

Advancing the cam position begins the intake event sooner, opens the intake sooner, builds more low-end torque, decreases piston-to-intake valve clearance, and increases piston-to-exhaust valve clearance. Conversely, retarding the cam position delays the intake event and opens the intake valve later, builds more high-end power, increases piston-to-intake valve clearance, and decreases piston-to-exhaust valve clearance.

Factory Camshafts

Listed here are specifications for stock production camshafts in Gen IV engines. All factory LS engines feature hydraulic roller lifters and roller camshafts.

GM Camshaft Specifications				
Engine	GM PN	Duration at .050	Lift	LSA
LS2	12574519	204 intake/211 exhaust	525/525	117
LS7	12638426	211 intake/230 exhaust	558/558	121
LS3	12603844	204 intake/211 exhaust	551/522	117.5
LS9	12638427	211 intake/230 exhaust	562/562	122.5

Active Fuel Management (AFM)

Some LS engines feature active fuel management (AFM) involving either displacement on demand (DOD) or variable valve timing (VVT). DOD essentially shuts off specific cylinders via oil-pressure/spring-controlled/spring-assist special lifters when engine demand is low. VVT works by adjusting valve timing depending on engine RPM: retarding valve timing at high RPM and advancing timing at low RPM. The goal of either system is, in theory, to gain fuel economy. In reality, for a high-performance build, it's best to delete these systems to obtain maximum power potential. To delete an active fuel management system and take advantage of a high-performance aftermarket camshaft, active fuel management delete kits are readily available, both for DOD and VVT systems.

DOD System

A DOD system disables half of the cylinders (1-4-6-7) under cruising or low-load conditions by collapsing the lifters in those cylinder locations and by cutting the spark for the same cylinders. When demand increases, the lifters are activated and "normal" lifter performance returns. The system is operated by the engine control unit (ECU) and four solenoids

DOD lifters feature "assist" springs. When the engine is under light load, the ECU signals a cutoff of oil supply to the select DOD lifters, allowing those lifters to collapse so that the intake and exhaust valves in the DOD locations close, shutting off firing to the four select cylinders. The spring on the DOD lifters maintains just enough pressure to the pushrod to prevent the pushrods from loosening and falling out. The DOD lifters feature a notched anti-rotation design that differs from standard LS lifters, requiring a different DOD-specific plastic roller guide tray.

A DOD delete kit, required for installing a non-DOD camshaft, includes a set of standard LS, non-DOD lifters, lifter guides, valley cover, new OEM-style head bolts, head gaskets, PVC hose, and PCV plug. This example is Trick Flow's PN 30678503. (Photo Courtesy Trick Flow Specialties)

DOD roller lifters retain the same .750-inch-diameter roller as standard LS roller lifters.

Displacement-on-Demand Delete

Parts needed to perform a DOD delete:

- Camshaft of choice
- Lifters of choice (OEM style or aftermarket-link-bar style)
- Lifter guide retainers (LS2 or LS7 style if using factory-style lifters)
- 4X three-bolt cam gear
- Cam gear bolts (ARP recommended)
- New camshaft retainer plate
- Valley cover (L92 style)
- ARP cam retainer plate screws
- New crank bolt (New OEM or ARP)
- Water pump gaskets
- Timing cover gasket
- Non-AFM valley cover, LS2/LS3 style
- Cylinder head gaskets (appropriate for your specific block)
- New cylinder head bolts (or studs)
- Front crank seal
- PCV hose and PCV plug

Note: I highly recommend using a new ARP crankshaft bolt instead of an OEM bolt. The factory bolt is a torque-to-yield style and should never be reused. Also, installation involves a torque-plus-angle method. The ARP bolt is reusable, and installation requires only a torque value. The old factory crank bolt may be used to press on the crank pulley, but never reinstalled for engine operation.

located in the upper valley, attached to the valley plate assembly known as a lifter lower manifold assembly (LLMA). The solenoids provide pressurized oil to the special roller lifters. The system is tuned to a specific factory camshaft, with the lobe profiles between the AFM and non-AFM cylinders different because of different valve lash requirements. Upgrading to a high-performance camshaft requires not only the new camshaft, but also replacing the special AFM lifters and lifter retainers, replacing the valley cover, eliminating the four solenoids, and plugging the oil delivery towers in the valley. In addition, the ECU must be reflashed/recalibrated. Plugging the oil towers is outlined in chapter 2.

While DOD or VVT factory cams might feature a single-bolt cam gear and cam, most aftermarket performance cams feature a three-bolt design. Because of this, part of deleting AFM involves obtaining a new 4X cam gear that features a three-bolt design.

Using a DOD camshaft along with a DOD delete kit won't work; it will result in a misfiring engine. A non-DOD camshaft is required.

VVT System

The VVT AFM system operates by the ECU monitoring valve timing through the camshaft position sensor. The ECU sends a signal to a solenoid that's mounted to the timing cover. The solenoid controls oil flow through a control valve built into the single camshaft bolt, operating a camshaft phaser actuator on the timing cover that rotates to advance or retard the camshaft.

VVT was first employed in select 6.0L and 6.2L LS engines. The front-mounted actuator controls the amount of intake and exhaust valve overlap. The engine's ECU commands the actuator to advance or retard the camshaft. The VVT camshaft features an internal oil passage that sends oil to the actuator. The actuator uses oil hydraulic pressure to change the

An example of a VVT delete kit from Trick Flow Specialties, PN 30678505. A new timing cover (type LS2/LS3) featuring the camshaft phaser actuator is included to replace the VVT cover. (Photo Courtesy Trick Flow Specialties)

camshaft timing. An actuator magnet is mounted to the front face of the timing cover, connected to 12 volts. When the solenoid is energized, the electromagnetic force on the magnet positions the spool valve of the solenoid. During engine operation, the ECU adjusts cam timing within a range of about 7 degrees of advance to about 55 degrees of retard.

The VVT system features a single-bolt camshaft. The cam bolt incorporates an internal spool valve with an internal check ball and filter. Pressurized oil that runs through the cam is sent through the spool valve to control the timing actuator. The valve spool directs oil out of advance/retard oil ports to the actuator.

The nose of the VVT cam bolt/spool valve features a face protrusion that contacts the magnet of the timing cover cam position actuator.

The VVT system doesn't like high-performance cams, which limit valve lift and duration. Also, higher valve spring pressures required for a high-performance cam can overwhelm the actuator. Just get rid of the VVT. Deleting a VVT system is slightly less intrusive or complex than performing a DOD delete. Deleting VVT only requires swapping out the cam gear, timing cover, cam position sensor, and cam gear bolts, in addition to the performance camshaft of choice. The ECU will also need to be recalibrated for non-VVT operation. A VVT delete kit will be necessary when removing the active fuel management VVT system from a Gen IV engine.

Typical delete kits include a three-bolt 4X cam timing gear, timing chain damper, LS2/LS3 timing cover, eight timing cover bolts, timing cover gasket, timing cover front seal, cam position sensor, cam sensor harness bracket, cam sensor harness, three ARP camshaft bolts, and water pump gaskets. The cam sensor features a three-pin connector, as opposed to the five-pin VVT connector. To use the new sensor, you need to remove the wires from the three-pin connector and reuse the

The cam position actuator magnet mounts to the exterior of the timing cover. The power connector is positioned at about seven o'clock when mounted.

The rear of the cam position actuator magnet places the electromagnet facing the VVT cam bolt/spool valve nose.

Variable Valve Timing Delete

Parts needed to perform a VVT delete:

- Camshaft of choice
- 4X three-bolt cam gear (LS2/LS3 style)
- ARP cam gear bolts
- Timing cover (LS2/LS3 style)
- Camshaft position sensor (LS2/LS3 style)
- Timing cover gasket
- Water pump gaskets
- New cam retainer plate
- ARP cam retainer plate bolts
- Crank bolt (OEM or ARP)
- Front crank seal

A VVT camshaft features an internal oil passage that allows pressurized oil to be delivered to the cam bolt/spool valve. The majority of aftermarket performance camshafts feature a three-bolt design, so a three-bolt cam timing sprocket is needed to accommodate the new performance cam.

original five-pin connector, or you can simply swap the cam sensor from the existing VVT cover to the new cover. Some kits also include a new camshaft retainer plate and crankshaft bolt.

Gen IV LS3 and LS7 engine packages offer tremendous power potential due to the larger displacement and superior-flowing cylinder heads, but the weak spot involves the camshaft. Factory cams are designed to provide a compromise between power, torque, idle quality, and fuel mileage. Substantial power gains can be had by simply changing cam profile, by as much as a whopping 70 hp. Keep in mind that factory valve springs are mated to the factory cams. When increasing valve lift and extending the RPM range, higher-rated valve springs must accompany a cam swap.

As far as AFM systems are concerned, whether DOD or VVT, just eliminate this nonsense. Each system involves added components that increase the potential for parts failures, and they limit performance potential. If you're after optimum performance for the street or track, just delete the AFM and get back to basics that allow you to take advantage of LS power potential. However, if your primary goal is to improve fuel economy instead of maximizing power, just buy a Prius hybrid and blend into the crowd of non-car folks.

Intake and Exhaust Difference

While some cam makers choose to use the same lift at intake and exhaust locations, some prefer to use a bit less lift at the exhaust to promote better exhaust scavenging. Exhaust valves are smaller than intake valves, so we have less exhaust volume than intake volume. Exhaust exits the cylinder head based on piston displacement, cylinder pressure, and header pull-out scavenging. By running less lift at the exhaust valve than at the intake, the exhaust can speed up, creating more vacuum and exiting faster. If exhaust volume is too large, it can reverse, which is detrimental to power.

The use of electronic fuel injection (EFI) in LS engine applications is obviously commonplace, but if you decide to run a carburetor setup, you have two basic choices in terms of duration and lobe separation. If you run a carburetor with a cam that has a shorter duration and tighter lobe separation, the engine will be "snappier," building peak power more quickly but also losing it more quickly. However, by running a longer duration and wider lobe separation, you broaden the power curve for better midrange and top end.

Camshaft Firing Order

In certain racing applications, a special firing order camshaft (SFO) can be used as a *tuning* aid, allowing the competition engine builder to further address combustion heat and crankshaft disturbance (harmonic) issues. The goal in changing firing order is to create a smoother-running engine, more-even fuel distribution, and enhanced crankshaft and main bearing durability. In the process, horsepower gains may be achieved as well (no guarantees here, but in most cases a slight power increase does result).

The traditional firing order for Chevy small-block and big-block engines has always been 1-8-4-3-6-5-7-2, which is determined by the crankshaft rod pin layout. Each cylinder has a "companion" in the firing order. This companion cylinder will reach top dead center (TDC) at the same time as its counterpart, one on the power stroke and one on the exhaust stroke. These cylinders are paired as 1 and 6, 2 and 3, 4 and 7, and 5 and 8; these can be interchanged in the firing order without altering the crankshaft. Some builders report seeing no power improvement, and other builders claim to have achieved power gains by switching cylinders 4 and 7 to create a new firing order of 1-8-7-3-6-5-4-2. This can enhance fuel distribution, especially in open plenum–type intake manifolds, and reportedly can result in an added 5 to 10 hp.

For example, Pro Stock drag engines typically take advantage of a 4/7 swap. Swapping number-7 and number-4 cylinders in the firing order eliminates the fuel distribution and heat problems caused by cylinders number-5 and number-7 firing in succession. With the revised firing order, the two end cylinders do not have to fight for fuel from the manifold plenum. The result, in many cases, is a measurable power increase (typically 8 to 10 hp) and a smoother, cooler-running engine.

Playing with special firing orders isn't limited to the advanced race engine builder. General Motors adopted a special firing order in its Gen III and IV LS engine series, which feature a 4/7 and 2/3 swap for the same reasons: to smooth out the harmonics in the pursuit of greater engine durability and to potentially generate more power. In the early development of the LS, GM's analysis showed that main journal number-4 had peak leads that were significantly higher than in main number-2. By changing the firing order to 1-8-7-2-6-5-4-3, the peak loading on main number-4 was reduced and the peak loading on number-2 went up. Overall, the loading among the main journal locations showed improved balance. In turn, the oil film between main bearings and main journals was better balanced at all main locations.

The primary reason to alter cylinder firing order is to achieve a smoother-running engine that delivers a more lineal acceleration ramp with less harmonic effect and crank deflection on the crankshaft and its main bearings. The goal of swapping firing order positions is to reduce crankshaft harmonic effects caused by two adjacent cylinders firing in succession (companion cylinders). By strategically relocating these companions, it's possible for the engine to idle and run smoother, to reduce isolated hot spots (cylinder-to-adjacent-cylinder walls), and to even out fuel distribution, primarily in applications that feature a single-plane intake manifold, and even more noticeably in tunnel ram intake manifold applications.

With a fuel injection setup, a lean cylinder can be richened (via the engine controller) to eliminate detonation, so firing order changes may not be as beneficial in an injected engine because fuel is delivered on an individual-cylinder basis. However, the LS firing order takes advantage of a 7/4 and 3/2 swap as an additional tuning aid, to produce an even smoother acceleration profile and to benefit crank and bearing life.

Firing Order Examples: Small-Block/ Big-Block Chevy	
Standard firing order	1-8-4-3-6-5-7-2
4/7 swap firing order	1-8-7-3-6-5-4-2
LS firing order (OEM GM)	1-8-7-2-6-5-4-3

Hydraulic or Solid

Hydraulic camshafts are designed to use hydraulic roller lifters. Once valve lash is set, no further lash adjustments should be needed. Hydraulic lifters provide smooth response and the hydraulic valve inside the lifter compensates for harmonics and for thermal expansion and contraction and dampens shock to the cam and valvetrain, making them ideal for street use.

Solid roller cams and lifters are intended for maximum performance and allow the use of more aggressive cam profiles. Solid cam and lifter combinations also better withstand higher valve spring pressures. When valve lash is set to the tight side of around .010 to .012 inch, solids also offer superior throttle response. The limitations, or drawbacks, if you will, involved with solid lifters include a higher price tag; the need for periodic valve lash adjustment due to variations of heat and thermal dynamics; increased noise, especially during cold starts; and poor suitability for extended low-RPM operation. Yes, a solid roller cam can be used in street applications, but routine checking and adjustment of valve lash is mandatory, something that many street enthusiasts may not be willing to perform. In short, for a street application, go with hydraulic. For racing use, go with solids. Aftermarket cam makers offer a wide range of profiles for LS applications in both hydraulic and solid formats.

Examples of extreme-duty race-level lifters include those offered by Comp, Lunati, Morel, and Gatorman (formerly Crane, available through Howards Cams), and for extreme-duty solid rollers, a brand called BAM.

Cam Gears

All Gen IV camshaft gears feature a 4X reluctor lug pickup design, for front-mounted camshaft position sensor location. On Gen III engines the camshaft position sensor was mounted at the top rear of the block. This is an easy way to identify a Gen III versus a Gen IV engine.

All Gen IV cam gears mount to the camshaft with a three-bolt attachment. The exceptions are the 2007 LS2 and 2008 and later LS3 and LS9 applications that feature a single

Gen IV cam gears feature a three-bolt hole pattern for cam attachment except the 2007 LS2 and 2008 LS3 and LS9, which feature a single-bolt mount.

The three-bolt Gen IV cam gear, featuring a 4X lug for Gen IV camshaft position sensor pickup, is available as PN 12586481 (shown at left). Three-bolt 1X cam gears, PN 12576407, were used only in 2005 LS2 Corvette and 2005–2006 LS2 GTO/SSR applications.

The three-bolt 4X cam gear is designed for a front-of-block-mounted camshaft position sensor. Note the timing notch, which aligns at six o'clock in relation to the crank gear at twelve o'clock, with the number-1 piston at TDC with the number-1 valves closed. All LS cam gears feature this timing reference notch.

An easy way to identify a Gen III versus a Gen IV is the location of the camshaft position sensor. If the sensor is mounted at the timing cover, it's a Gen IV engine.

Aftermarket performance cam makers offer LS cam kits that include either OEM-style LS roller lifters, as shown here, or lifters that are tied together in pairs via a link bar. The LS lifters that copy the factory lifter dimensions and basic design allow the use of the original plastic lifter guides, whereas linked lifters eliminate the need for the plastic guides. Aftermarket cam makers such as Crane, Lunati, Comp, Bullet, and others offer an extremely wide selection of cam profiles for any LS application, from mild street to extreme race applications.

Camshaft and Cam Gear Designs

GM PN 12591689	Single-bolt 4X cam gear for 2007 LS2 and 2008 LS3/LS9
GM PN 12586481	Three-bolt 4X cam gear for 2006 LS2 Corvette and 2006–2009 LS7 Corvette
GM PN 12576407	Three-bolt 1X cam gear for 2005 LS2 Corvette and 2005–2006 LS2 GTO/SSR

center-bolt gear-to-cam mounting. However, aftermarket cams and cam gears are available for LS3 and LS9 applications that feature a three-bolt design.

All Gen IV engines feature the camshaft position sensor up front, at the timing cover. The camshaft position sensor part number for Gen IV 4X cam gears is 12591720. This cam position sensor applies to LS2, LS7, LS3, and LS9 applications. If you need a camshaft position sensor harness (doing a swap, etc.), the GM part number is 12627501.

Lifters and Lifter Guides

In an LS engine build, you have two choices of lifters: OEM or aftermarket. OEM-style lifters must be guided within an OEM plastic guide. Aftermarket performance lifters are connected in pairs via a pivoting tie-bar link.

Factory plastic lifter guides provide a register feature: the lifters have opposing flats on the upper body that register into flats in the plastic guides. This style is certainly adequate to maintain the lifter roller bearings in plane with the cam lobes. While flat-tappet lifters are designed to rotate in their bores during operation, roller lifters must remain in a fixed plane to allow the roller bearings to roll against the cam lobes. If the lifter rotates a bit, the rollers may tend to skip and chatter along with cam lobes. A superior approach is to use high-quality aftermarket roller lifters that are tied together in pairs with a pivoting link bar. This style provides a much more precise design that keeps the lifter rollers in plane with the cam lobes.

Over time, age and wear to the plastic guides may allow the lifters to slightly rotate out of plane; perhaps not enough to destroy a cam but enough to reduce efficiency and the long-term durability of both the lifter roller and the cam lobe. For maximum performance and durability, linked tie-bar lifters are a superior choice, especially when you're chasing maximum performance and sustained high engine speeds. In short, if you're planning to slam the engine for all it's worth, get rid of the factory plastic guides and factory-style lifters and upgrade to tie-bar roller lifters.

If you're using the OEM plastic lifter guides, be aware that excess oil tends to puddle at the floor of each lifter's location. For faster oil drainback, drill about a 5/16-inch diameter hole at the outboard side of the guide, just above the floor at each lifter location. The outboard side is the one that faces the exhaust side of the engine. After drilling the holes, carefully deburr to remove any plastic fragments, then wash and clean to make sure that the guide is free of debris.

If you're using aftermarket roller lifters that are connected in pairs with a tie bar, you have no need for the OEM plastic lifter guides. The purpose of using OEM guides or tie-bar lifters is to maintain the lifter rollers in plane with the camshaft lobes. Unlike flat-tappet lifters that are designed to rotate during operation, roller lifters must be locked in plane to prevent rotation, allowing the lifter roller tips to roll against the lobes. If a roller lifter rotates in its bore, the lifter will crash and scrub against the lobe, resulting in very quick and catastrophic damage to both the lifter and cam lobe.

Note that due to the design of the LS block and cylinder heads, lifters may be installed or removed only with the cylinder heads and head gaskets removed. Unlike early generation small-block engines that allow lifter access by removing the intake manifold, gaining access to LS lifters requires removal of the heads and head gaskets.

High-performance roller lifters are available individually to use with OEM plastic lifter guides or as tie-bar-connected pairs that eliminate the need for the plastic lifter guides. The only downside of using tie-bar roller lifters is that during a camshaft change, the lifters must be removed in order to reach the cam. The plastic lifter guides allow you to pop the lifters up and away from the cam lobes without the need to remove the lifters. With the rocker arms and pushrods removed, by rotating the crankshaft twice, the lifters are pushed up into the "locking" position where they are slightly gripped by the guides, holding the lifters up and out of the way. However, not having that feature is a small price to pay for the advantage of the extremely durable tie-bar roller lifters that are available in today's aftermarket.

When securing the OEM plastic lifter guides, the OEM 6-mm x 1.0 bolts must be used; they feature a shoulder under the bolt head that properly registers the lifter guide to the block. These bolts are tightened at 106 in-lbs. Applying a drop of Loctite 242 blue thread locker isn't a bad idea.

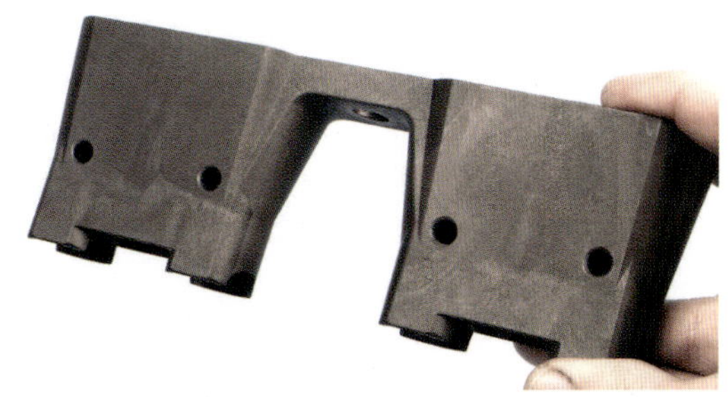

Lifters guides, also referred to as lifter buckets or lifter trays, tend to collect excess oil at the outboard sides. It's a good idea to drill a 5/16-inch hole at the lower outboard side of each bank to allow better drainback.

While factory LS cams may feature a single-bolt or three-bolt connection of the cam gear to the cam, the majority of aftermarket performance cams feature the three-bolt design.

Pictured here is a factory roller lifter from a 5.3L engine (left) and an aftermarket roller lifter for the same application. The performance lifter features a more robust roller bearing and heavy-duty bearing axle. Also notice the difference in overall length.

Install the camshaft retainer plate with new screws. ARP retainer plate screws are highly recommended. Apply a drop of medium-strength thread locker to each screw's threads, and tighten to 18 ft-lbs.

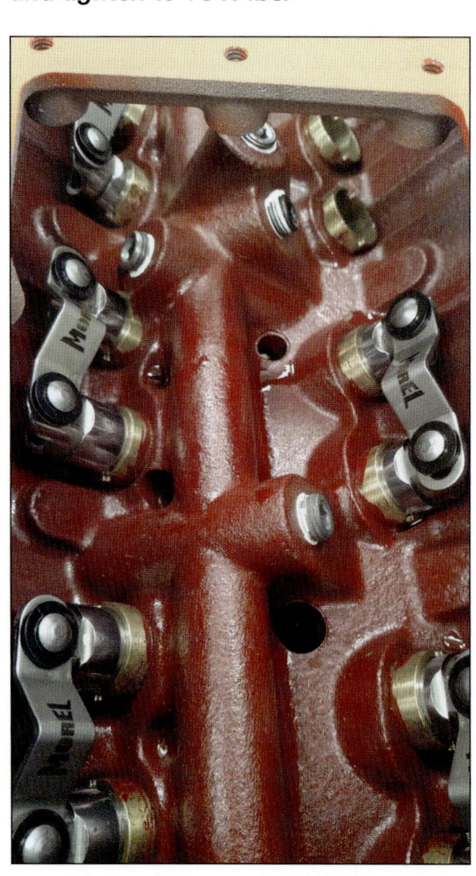

An aftermarket lifter may feature a deeper pushrod cup location. This is one more example of why it's important to measure for pushrod length instead of assuming that factory-length pushrods will suffice.

Roller lifters that feature a tie-bar link are available with either a straight link or a V-shaped link. The V-shaped link is available to provide added clearance between the link and the block's bosses in the lifter valley.

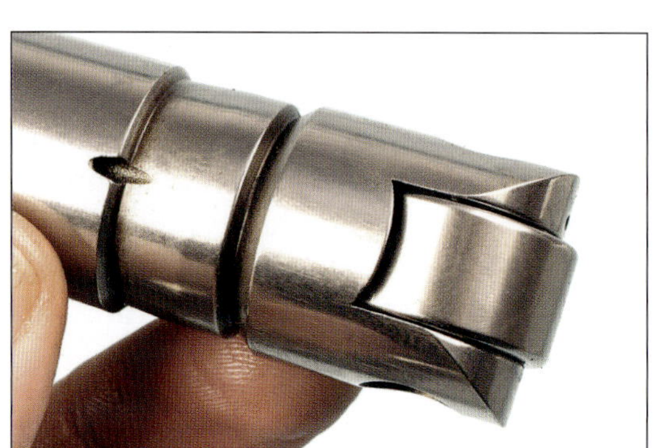

High-quality aftermarket performance roller lifters feature stronger roller bearings and are available with larger-diameter rollers that help to increase duration. High-performance lifter rollers are designed to accommodate higher valve spring pressures and higher engine speeds while maintaining durability. The example shown here is a roller lifter from Morel, a firm that also supplies roller lifters to several performance aftermarket cam and valvetrain manufacturers.

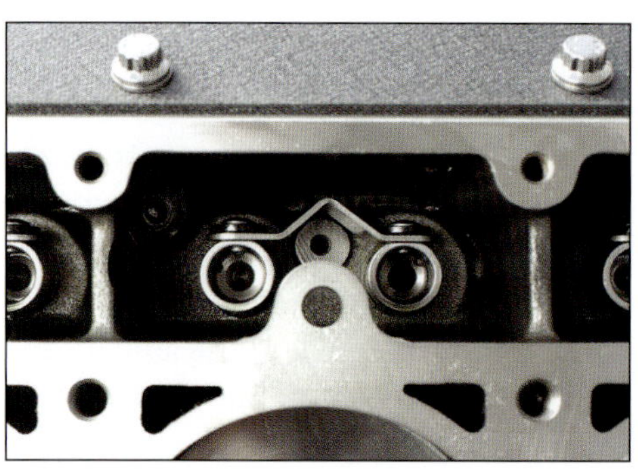

Depending on the height of the boss, a V-style tie-bar link may be necessary to provide clearance.

Roller lifters that are connected by a tie-bar link are able to cycle vertically in their bores, and the links are better than the OEM plastic lifter guides at preventing lifter rotation.

Many performance builders prefer bronze lifter bushings, which provide superior lubricity for the lifter bodies. Installation requires overboring the lifter bores, press-fitting the bronze bushings, and machining the bushing inside diameters to provide lifter bore oil clearance, which is usually in the .0015-inch to .0018-inch range. Machining the lifter bores and sizing the bushing inside diameters should not be performed with abrasive honing stones, as this may result in inconsistent inside diameters along the length of the bushings. The preferred method is to machine with the appropriate cutters, either using a specialty fixture to obtain the correct geometry and bore angle or with the use of a CNC-machining center. This photo shows bronze lifter bushings installed in a World Products Motown II LS block, but the same practice applies to conventional LS blocks.

Rocker Arms

In production applications, all Gen III rocker arms feature a 1.7:1 ratio. All Gen IV rocker arms also feature a 1.7:1 ratio, except LS7 heads, which feature a 1.8:1 ratio. On Gen IV LS2 engines, as with Gen III, the intake and exhaust rocker arms are identical, with valve tip and pushrod cup inline.

Gen IV LS3, LS7, LS9, and L92 heads feature offset *intake* rockers to accommodate larger intake valves. The exhaust rockers are the same for all LS engines except LS7 applications, which feature a 1.8:1 arm ratio. All Gen IV exhaust rockers are interchangeable except for the LS7 because of that application's higher 1.8:1 ratio. All Gen IV intake rockers are interchangeable except for the LS2 intake rockers, which are not offset. In summary, intake rockers for LS7, LS3, LS9, and L92 are offset, with the only difference being the longer LS7 arm ratio.

GM Gen IV Rocker Arm Part Numbers			
Cylinder Head	Intake	Exhaust	Ratio
LS2	10214664	10214664	1.7:1
LS7	12579615 (offset)	12579617	1.8:1
LS3/LS9/L92	12569167 (offset)	10214664	1.7:1

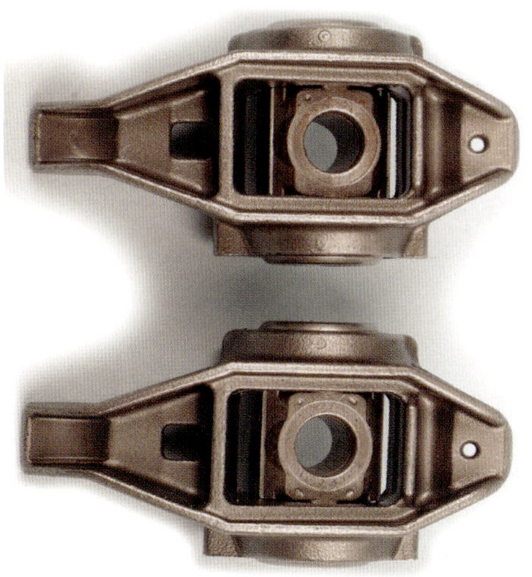

The intake and exhaust locations on these LS2 rockers are identical, with no offset at the intake rocker, so they are interchangeable for location. The factory ratio is 1.7:1, and these same rockers were used on earlier Gen III LS heads.

LS3, LS9, and L92 rockers feature a ratio of 1.7:1 and offset intake rockers. Shown here is an intake rocker (top) and exhaust rocker (bottom). The exhaust rockers are identical to the intake and exhaust rockers for the LS2.

LS7 rocker arms feature the same intake offset configuration as LS3, LS9, and L92 rockers, but the ratio was increased for LS7 applications to 1.8:1.

These two exhaust rockers provide a visual comparison: If you look closely you'll note the increased length of the LS7 rocker. The rocker at the top features a ratio of 1.7:1 (applicable to LS2 intake and exhaust, and LS3, LS9, and L92 exhaust locations). The rocker at the bottom is an LS7 rocker with 1.8:1 ratio.

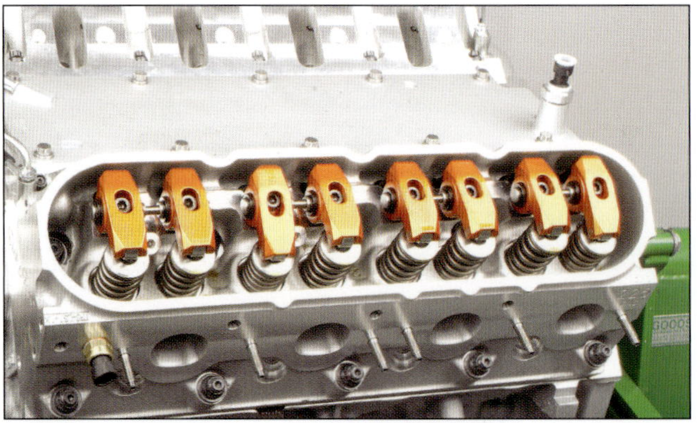

This LS2 rocker setup features Comp aftermarket forged aluminum rockers. Notice that all intake and exhaust rockers are straight, with offset at the intake positions. These rockers are interchangeable for intake or exhaust positions.

This close-up shows the flat machined bosses on Gen IV heads (except LS7) that accept a cast-aluminum rocker mounting rail. The radius on the rail features a pocket for the rocket trunnions.

LS7 cylinder heads (as well as some aftermarket heads) feature radiused pedestal stands that directly accept the rockers, with no need for a rocker mounting rail.

Some Gen IV heads, such as this Trick Flow LS3-style head, feature a flat-milled rocker mounting base area. This allows the use of flat-faced mounting pedestals or a rail with a flat underside and radiused pedestals to permit the use of two different styles of rocker trunnions. OEM rockers with a round trunnion require rail adapters, while rockers with flat-faced trunnions require individual flat-faced pedestal stands.

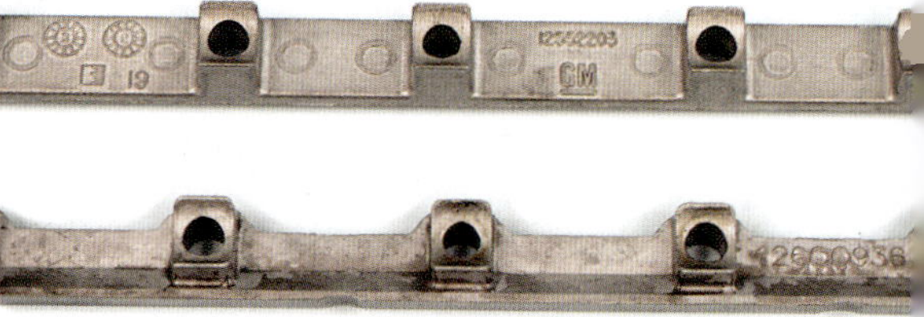

Shown at the top right is a rocker rail for Gen III and LS2 Gen IV heads. The rail at the bottom applies to LS3, LS9, and L92 heads and features different spacing for offset intake rockers.

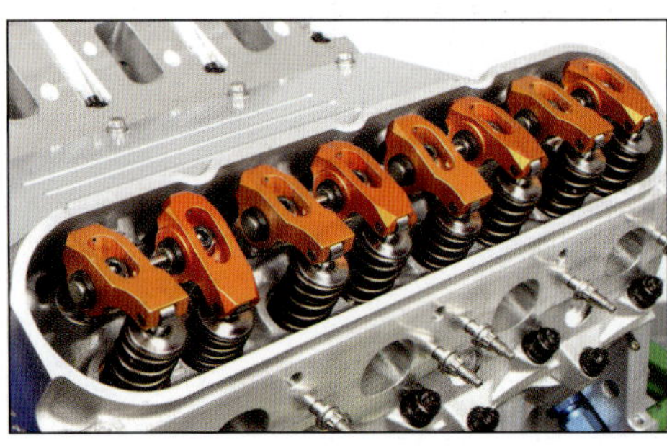

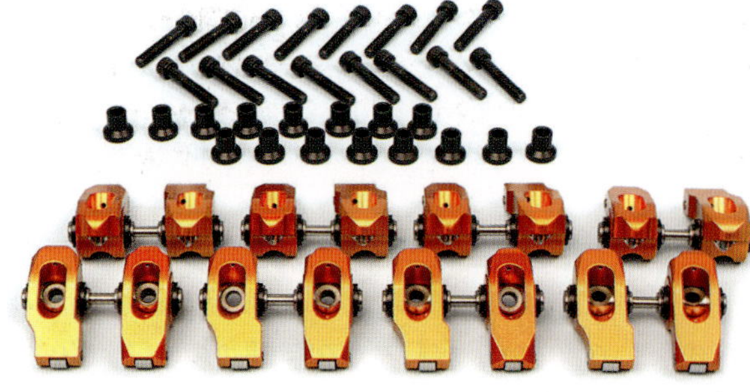

This set of Comp Cams rockers features offset intake rockers to accommodate LS3- or LS9-style heads with a flat-milled mounting base. The set includes short mounting pedestals and 8-mm socket-head cap screws.

All OEM-style LS rocker bolts for OEM and aftermarket aluminum heads feature 8-mm-diameter threads and are torqued to 22 ft-lbs.

Stock LS Rocker Arm Trunnion Upgrades

Rocker arms are a necessary upgrade for an LS weak link. All GM factory rocker arms are manufactured of powdered metal, which is a surprisingly durable pressure-cast material, generally suitable for horsepower ratings up to about 500 hp. For higher-horsepower builds, performance aftermarket stronger forged/billet roller rocker arms are strongly recommended, especially when you're upgrading to a performance cam and higher-rate valve springs.

The GM factory rocker arm's weak areas are the trunnion shaft and the trunnion bearings that the rocker arms pivot upon. The OEM trunnions are made of powdered metal and feature uncaged needle bearings. These have been known to separate at sustained high RPM, resulting in the bearings separating and scattering through the engine. While the factory rocker arms are strong enough for spirited street driving, it is highly recommended to upgrade all rockers with Comp Cams' trunnion upgrade kit. It features high-strength 8620 steel-alloy trunnions and caged roller bearings.

For performance use, get rid of the OEM powdered metal trunnion and the cheap uncaged bearings in favor of a durable aftermarket trunnion kit upgrade. The OEM setup is shown here. Notice the small bearing needles? How'd you like to have a bunch of these floating through your engine?

The OEM trunnion bearings are captured in a two-piece bearing housing. The outer half of this tin housing tends to walk out, releasing the bearing needles.

Swapping out the trunnions and bearings is an easy task that can be handled by any skilled do-it-yourselfer, requiring only press adapters and a bench vise. Don't even debate the issue. If you plan to use GM rocker arms, go ahead and install the upgraded trunnion kit. This will eliminate the concern. The Comp Cams trunnion upgrade kit is available as PN 13702-KIT. This is an essential upgrade for any factory LS rocker arm set.

The recommendation to upgrade LS rocker arm trunnions is not limited to performance enthusiasts alone. Any LS engine that will be exposed to abusive conditions will benefit from the upgrade. This can include LS engine applications for police vehicles and other emergency vehicles, as well as commercially used trucks that experience heavy engine loads.

If, during an oil change, you find small "mystery" needle bearings in the oil or stuck to the magnetic drain plug, this is clear evidence of rocker arm needle bearing failure. The needles have left the rockers and are now wiggling around on top of the heads and washing down through oil drainback holes on their way to the sump. And along the way, who knows what little orifices they'll migrate into and cause damage? As I mentioned earlier, this condition is more likely when the engine is frequently operated at high RPM and/or high valve spring pressures are used, as when changing over to a hotter cam that includes higher-rate springs. This warning bears repeating. The inexpensive OEM rocker arm bearing setup is a proven weak link in the LS system.

Knowing that this possibility exists, the best course of action is to address the issue *before* it becomes a problem. You need to change to high-quality aftermarket performance full-roller rockers from any reputable aftermarket rocker manufacturer because they use high-quality caged bearings in their rockers. Or you need to upgrade the original rocker arms. Especially for those on a limited budget, upgrading the OEM rockers is an excellent option.

Comp Cams, as an example, offers a very high-quality trunnion bearing upgrade kit, PN 13702-KIT. This kit is made for LS1, LS2, LS3,

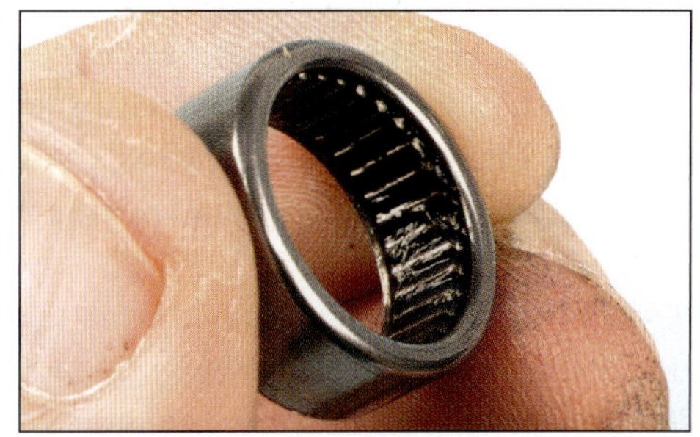

Comp's fully caged roller bearings are a sensible upgrade from the uncaged OEM bearing assembly.

Before performing the upgrade, inspect your OEM rockers. If the pushrod cups appear worn, as on this example, trash the rocker. OEM rockers are made of investment-cast case-hardened steel. If the cup is visibly worn, the case hardening has been compromised, and the cup will continue to wear rapidly.

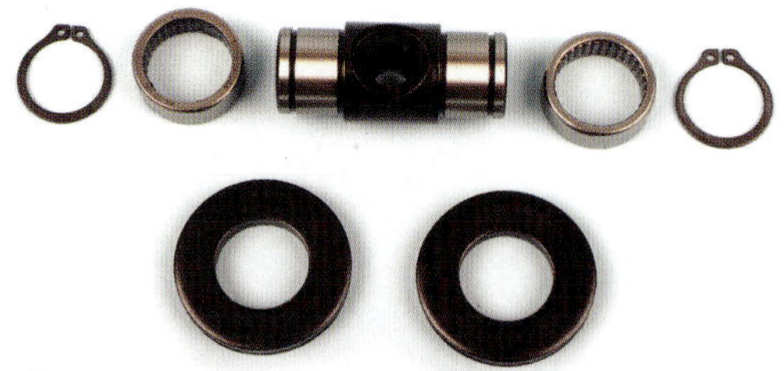

The Comp upgrade kit features high-strength steel trunnions, caged bearings, and peace-of-mind circlips. There's no way this setup will ever allow bearing needles to fall out. The Comp upgrade kit also includes a pair of thick steel washers that are used during the press-fit of the new bearings.

This photo is for illustrative purposes. Here, a pair of bearings are placed onto the new trunnion. Notice the circlip groove at each end. In addition to the interference fit of the bearing cages into the rocker arm, the additional security of circlips ensures positive capture of the bearing cages. During assembly, the kit's washers provide positive stops during the press-in of the bearings.

The rear of the tool base (facing the bench vise) features two small, very strong magnet buttons. This holds the tool base against the vise jaw.

LS6, and LS7 OEM rockers. It includes new trunnions, new caged and pre-lubed bearing assemblies, circlips and rocker arm mounting socket head cap screws, as well as two thick installation-assist washers. The new trunnions feature extended tips with circlip grooves. Instead of the bearing housings relying on a press fit, the additional circlips serve to prevent possible bearing assembly walkout. These bearings aren't going anywhere, so the rocker bearing issue is no longer a concern. The new trunnions and bearings are also made of higher-quality materials and are designed to withstand higher RPM and high spring pressures. Comp's upgrade kit replaces the (cheap) cageless OEM needle bearing assemblies and powdered metal trunnions with very high-strength premium 8620 steel-alloy trunnions and fully caged roller bearings for *vastly* improved durability.

Tool Note

In addition to a bench vise, you'll need adapters that allow you to press out the old bearings and press in the new bearings. Comp's 13702TL-KIT includes two washers that serve as press adapters. An alternative is a specialty tool kit designed for this task, such as Summit Racing's PN SME-906011, which makes the job much easier. If you need a set of new GM rocker bodies, Summit also offers an eight-pack of brand-new, shot-peened GM rocker bodies (without bearings, so there's no need to press out the old ones). These rockers are available under Summit's PN SME-143000.

Summit Racing's new trunnion vise tool, PN SME-906011, allows you to easily and safely remove the original bearings and trunnions and install the upgraded Comp Cams trunnion kit using a common bench vise instead of a hydraulic press. Removing old trunnions and bearings and installing the new Comp Cams upgrade kit in all 16 rockers can be done in as little as 30 to 45 minutes without breaking a sweat.

Trunnion Upgrade Procedure

Performing this retrofit upgrade is relatively simple, requiring no modifications to the rocker arm. Follow the steps below.

By the way, this is a good time to clean the rocker arm bodies. If you want the rockers to look factory fresh, clean them with solvent, have them *professionally* tumble cleaned with a fine-grit stone, or glass-bead blast the disassembled bare rocker bodies. If you do decide to blast with fine glass bead, be careful not to concentrate and dwell any blasting in the trunnion bores or on the valve tips or pushrod cups. *Do not* run any type of hone or reamer in the trunnion bores. If trunnion bores appear to be damaged (burred, scored, or pitted), replace the rocker arm. You must avoid altering the trunnion bore dimensions.

Using the Summit kit's presser (black piece) push the bearings and trunnion out of the rocker. The presser features a shallow counterbore that aligns to the stock bearing. Summit's specialty fixture on a bench vise is shown here. Push the old trunnion completely out of the rocker along with both OEM bearings.

Using the Summit kit's mandrel, place the new bearing onto the mandrel, followed by the aluminum alignment sleeve. This aligns the new bearing with the rocker arm body. Push a new caged bearing into one side of the rocker.

1 Place the Summit tool vise fixture onto one jaw of your bench vise. Place the OEM rocker into the fixture's V-saddle.

2 Align the (black) Summit presser to the OEM bearing, with the counterbored end of the presser facing the bearing. Tighten the bench vise to press out both OEM bearings and the trunnion (needles will fall out). Inspect/clean rockers as needed.

3 Place the (black) presser onto the short end of the Summit kit's mandrel. Slide a new bearing onto the long end of the mandrel (make sure that the stamped symbols on the bearing face the mandrel's flange).

4 Slip the aluminum alignment sleeve onto the long end of the mandrel, seated against the bearing.

5 Slide the mandrel assembly into the rocker arm bearing bore (the aluminum alignment sleeve enters the bore first to align to the rocker body).

6 Using the bench vise, press the bearing into the rocker body until the mandrel's flange bottoms out onto the rocker body. Perform this single-bearing installation to all 16 rockers before proceeding with trunnion and opposite-side bearing installation.

7 Place the rocker arm body onto the Summit vise fixture, with the previously installed bearing facing the fixture.

8 Slide the new trunnion halfway into the bearing bore (leaving the outside end of the trunnion exposed to align the remaining bearing). Apply engine oil to the outside of the bearing. Slip the remaining bearing onto the exposed trunnion (make sure that the stamped symbols on the bearing cage face outward).

9 Place the kit's thick steel washer onto the end of the trunnion (against the bearing face).

10 Tighten the bench vise until the washer bottoms out against the rocker body.

11 Verify that the trunnion rotates freely, and that a slight amount of endplay (side-to-side) exists, in the range of .004 to .008 inch. If you don't have any endplay, adjust bearing depth accordingly.

12 Using snap-ring pliers, install one circlip to each end of the trunnion. Verify that the clip is fully seated in the trunnion clip groove.

13 Install the upgraded rocker arms to the cylinder heads using the new socket head cap screws supplied in the Comp Cams kit. Be sure to rotate the trunnion so that the flat recess side of the trunnion's bolt hole faces upward (the socket head cap screw must seat against this recessed flat side of the hole). Tighten the cap screw to a value of 22 ft-lbs, following the OEM installation procedure.

Once one bearing is in place, insert a new trunnion (I suggest lubricating the trunnion first).

Place one of the kit's washers over the exposed end of the trunnion, against the outer face of the new bearing.

Before installing the circlips, verify that the trunnion rotates freely and that a small amount of endplay is present (.004- to .008-inch endplay). Using snap-ring pliers, install one circlip into each end of the trunnion. Rotate the trunnion and inspect to verify that each circlip is fully seated.

Insert the new trunnion so that one end engages the newly installed bearing. With the new trunnion in place, insert the next bearing onto the trunnion (this slips on). Make sure that all bearings are installed with the stamped lettering side facing outward!

Using a bench vise with Summit's specialty fixture to secure the opposite side, press the remaining bearing into the rocker, so that the outer face of the bearing is flush with the outer surface of the rocker body. Apply a bit of engine oil to the outside of the new bearings to ease press-in.

This close-up shows a circlip installed. Notice the small stamped letters on the outer surface of the bearing cage. Remember, the stamped letters of each bearing must face outward. This rocker is done and ready for high-performance use. Comp's bearings are fully caged and of the highest quality, and the machined steel trunnions offer much greater strength compared to the powdered metal OEM trunnions. Nothing against powdered metal, but 8620 steel is stronger and is more suited for higher valve spring pressures.

The second bearing has been installed on the new trunnion.

This overhead view shows a fully assembled rocker with Comp's trunnion upgrade. Notice that one side of the trunnion mounting hole is recessed (countersunk with a flat surface). This recessed side of the hole must face upward during rocker installation (the opposite side of the trunnion is round, to seat onto the rocker stanchion).

Depending on the version of your LS cylinder heads, some have integral rocker pedestals while others require separate pedestal rails. This 5.3L LS features rocker rails. Hand-snug a couple of rocker bolts to the center the rail before fully tightening. Also, be sure to apply high-pressure assembly lube to all valve tips, rocker pushrod cups, and rocker valve tips.

Instead of using the OEM hex-head rocker mounting bolts, Comp provides a full set of 8-mm socket head cap screws that require a 6-mm hex bit. The cap screw head nestles into the trunnion hole's recess. Apply a bit of assembly lube under the screw head before installation, and torque to Comp's recommendation of 22 ft-lbs.

Using a speed wrench with a 6-mm socket or 6-mm T-handle hex wrench, snug all rocker bolts.

Available Trunnion/Bearing Upgrade Items

Item	Mfg.	PN
Rocker arm trunnion upgrade	Comp Cams	13702-KIT
Rocker arm vise tool	Summit Racing	SME-906011
New shot-peened GM rocker bodies (without BRGS)	Summit Racing	SME-143000 (eight-pack)

Note: Install the new upgrade kit using either a press or with Summit's new rocker arm vise tool. It's critical to smoothly *draw* the OEM bearings out and *draw* the new bearings in. ***Do not*** *try to bang the old trunnions out or attempt to install the new bearings with a hammer.* Leave your hammer in the toolbox.

Aftermarket Rockers

The performance aftermarket offers upgraded rocker arms in both steel and aluminum, in ratios of both 1.7:1 and 1.8:1. For engines other than LS7, the stock ratio is 1.7:1; the LS7 rocker ratio is 1.8:1. If installing higher ratio 1.8:1 rockers on engines originally intended for 1.7:1 rockers, valve lift can be increased without a change to the camshaft.

For example, if the camshaft's lobe features a peak lift of .378 inch, when matched to 1.7:1 ratio rockers, effective valve lift would be .6426 inch. When the same cam is matched to rockers with a 1.8:1 ratio, valve lift is increased to .6804 inch.

Aftermarket rockers are also offered in a variety of mounting designs, depending on the cylinder head's rocker stand design, whether dealing with radiused-pocket stands as found on LS7 heads or flat stands as found on other LS heads. For heads that feature flat stands, rockers are available that require mounting plates with radiused pockets that allow the rocker trunnions to directly seat, or rocker systems with plate-based adapters that allow the pairing of two rocker arms (intake and exhaust) per adapter. Shaft-mounted rocker systems are also available, and these require flat mounting stands as part of the head casting.

For cylinder head styles such as LS7, LS3, L92, and LS9, rocker arm sets

Note the flat underside area of the performance aftermarket trunnions on Harland Sharp rockers, designed to mate to flat-faced pedestal stands with the use of standoff adapters supplied in the kit.

Harland Sharp LS1/LS2 roller rockers feature a tie-bar axle that pairs the intake and exhaust rockers, keeping both in plane.

that include straight exhaust rockers and offset intake rockers are readily available.

While OEM rocker arms are manufactured of powdered metal, upgrading to steel, billet, or extruded aluminum aftermarket rockers provides substantially greater durability under high-demand operation. Also, while OEM rockers feature a roller bearing trunnion pivot, the contact between the rocker arm and valve stem tip features a friction surface. Upgrading to full-roller rockers provides a drastic reduction of friction with the addition of a roller bearing at the rocker tip. If you're after a serious performance upgrade, switching to full-roller performance rockers adds durability and the ability to attain sustained higher engine speeds.

Depending on the mounting style, rockers are available as bolt-on, with no valve lash adjustment, or stud-mount, which provides valve lash adjustability. As you can see, the aftermarket offers a wide range of choices and versatility. As mentioned earlier in this chapter, if OEM rockers are used, it is essential to install a trunnion upgrade kit to eliminate the potential for OEM trunnion bearing failure.

Whenever upgrading to aftermarket roller rockers, be aware that thicker-body aluminum rockers may not clear OEM valve covers, in which case you'll need to add valve cover spacers or switch to taller valve covers. Always check for rocker-to-valve cover clearance at both valve closed and full lift positions. Rocker makers will usually let you know if their rockers require added clearance.

High-Performance Rocker Arms

Several performance rocker arms manufacturers currently offer roller lifters for LS engine applications, including but not limited to Crane, Comp Cams, Lunati, Bullet, Crower, Harland Sharp, Scorpion, Howards, Jesel, T&D, Yella Terra, Edelbrock, Erson, Ferrea, Isky, Manley, and Manton. Upgrading to aftermarket performance rocker arms provides several advantages. They can use full-roller designs that feature roller bearings at the rocker arm pivot point as well as a roller bearing for the arm-to-valve tip, which reduces friction. In addition, they offer stronger materials that better withstand high spring pressures and sustained high RPM. Add to this improved oil delivery to pushrods and valve tips and a choice of rocker

To utilize a shaft-mount rocker system, the rocker stands on the head must be flat, as found on LS3, LS9, and L92 heads. If you're dealing with an LS7 head, where the rocker pedestals feature a radiused pocket for the rockers, these can be milled flat to accept the shaft mount. (Photo Courtesy Comp Cams)

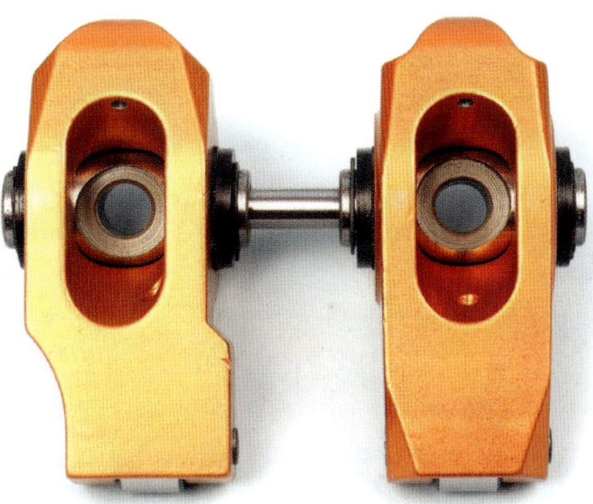

On this example of aftermarket Harland Sharp forged-aluminum rockers with offset intake, the trunnions are tied together with an axle that keeps the intake and exhaust rocker pair aligned and prevents clocking movement. This style of rocker system is available in 1.7:1 ratio for LS3 or LS9 applications, or 1.8:1 ratio for LS7 applications.

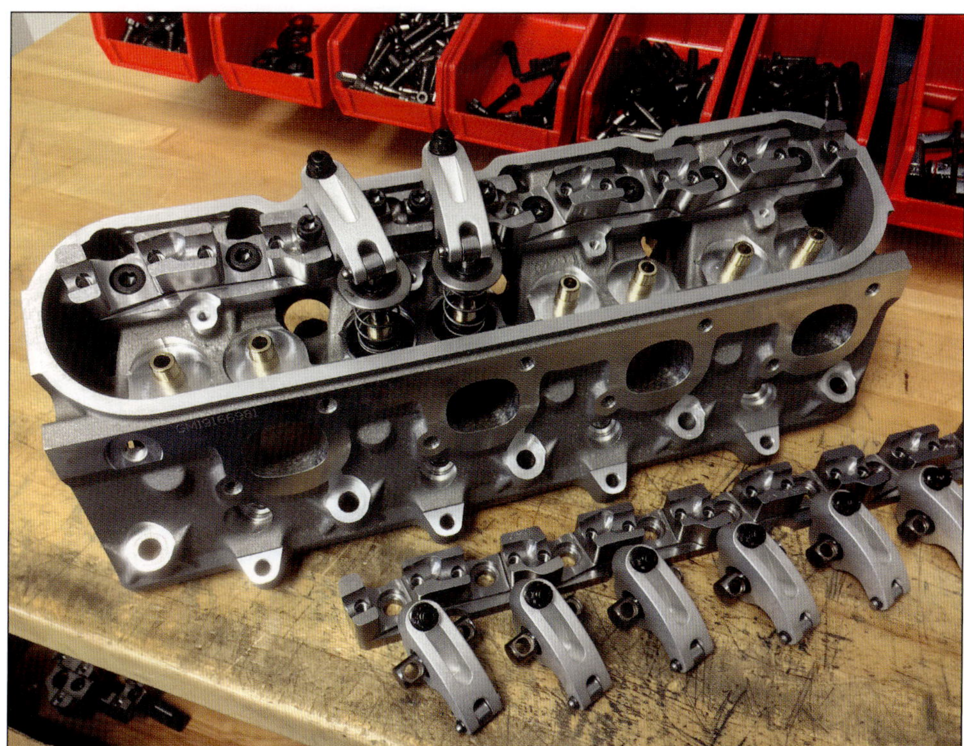

Shown here is an example of a Jesel rocker setup for an LS7 application with offset intake rockers. A mounting stand allows rocker installation. The aluminum rockers feature a standard-slot relief for weight savings. A variety of rocker profiles are offered. Note that these rockers are adjustable for valve lash. (Photo Courtesy Jesel)

arm ratios; for example, moving from the stock ratio of 1.7:1 to 1.8:1 increases effective lift with the same camshaft.

Comp Cams

Competition Cams offers three styles of LS rockers, including the Ultra Gold series, Ultra Pro Magnum series, and shaft-mounted rocker systems. The Ultra Gold series features CNC-machined extruded aluminum bodies, fully caged and high-strength trunnions, hardened steel pushrod cups, and roller bearing tips. This series is available in ratios of 1.7:1 or 1.8:1, with no offsets, which makes them compatible with LS1/LS2 heads. Each pair of intake and exhaust rockers shares a flat-bottom/radiused-pocket mounting stand. These rockers are nonadjustable bolt-ons, using 8-mm socket head cap screws provided.

The Ultra Pro Magnum series rockers are made of high-strength forged steel, available in either bolt-down/nonadjustable or stud-mount/adjustable versions. This series also offers a set that includes offset intake rockers for LS3/LS7 and LS9 applications. Ratio is 1.8:1.

Highly rigid shaft-mount rocker systems are offered for LS1/LS2 or LS7/LS3/LS9/L92 heads. The LS1/LS2 version is offered in either 1.7:1 or 1.8:1 ratio. The LS3/L92 version features .215-inch offset intake rockers, in 1.7:1 ratio. The LS7 version features offset intake rockers with a rocker ratio of 1.8:1.

Crower

Crower offers a wide selection of full-roller rocker arms and rocker systems for LS1/LS2 and all Gen IV L92, LS3, and LS7 applications, designed to fit OEM GM cylinder heads as well as rockers designed specifically for certain aftermarket heads, such as Brodix, Dart, Trick Flow, World Products, and more. Individual, pair-mount, and full-bank shaft-mount systems are available, as well as custom-order rockers.

Harland Sharp

Harland Sharp is renowned for its design and quality. Its LS offerings include forged CNC-machined aluminum bodies, heavy-duty trunnions with fully caged needle bearings, heavy-duty tip rollers, and hardened pushrod cups. Versions include direct replacement mounting on OEM heads for LS1/LS2 in

Note the intake rocker offset on this example of individual-mount steel roller rockers for an LS7 application. (Photo Courtesy T&D Machine)

This shaft-mount system is designed for LS7 applications. Shown here are solid-body aluminum rockers with offset intakes and a 1.8:1 ratio. (Photo Courtesy T&D Machine)

1.7:1 or 1.8:1 ratios, in bolt-on nonadjustable or adjustable design; L92/LS3 sets with offset intake rockers and tie-bar-connected intake/exhaust pairs in either 1.7:1 or 1.8:1 ratios, available as nonadjustable or adjustable versions; and LS7 1.8:1 rockers designed for direct replacement on OEM-style heads. The LS7 fully adjustable rockers require shorter pushrods and taller valve covers or valve cover spacers. Harland Sharp also offers aftermarket head-specific rocker systems for various cylinder head brands.

Howards Cams

Howards offers full-roller aluminum rockers designed as direct bolt-ons to replace stock rockers, with no need for valve cover spacers. Available for LS1/LS2, LS3/L92, and LS7. Available ratios include 1.7:1 and 1.8:1.

Jesel

Jesel has an extremely high reputation for race-quality rockers and will make just about anything you want in the custom shop. In addition it offers readily available rockers for a wide range of OEM and aftermarket heads. Its three basic families of rockers are the Sportsman Series, Pro Aluminum Series, and Pro Steel Series. All are available in shaft systems in addition to individual-mounted rockers. Applications include Gen III and Gen IV, designed for OEM GM heads, GM Performance, Edelbrock Performer RPM heads, Brodix, RHS, Trick Flow, Mast, World Warhawk, AFR Mongoose, Dart, and BMP heads. Rockers are available designed specifically for either hydraulic roller or solid roller cams. Ratios, depending on application, are offered in 1.7:1, 1.75:1, and 1.8:1.

Rocker bodies are available in solid, standard slot for weight reduction, or Mohawk beam for ultralight weight. Many custom options are offered, including optional needle bearing tips and optional ball adjusters that offer less friction than cup adjusters and eliminate the counterbore area for added strength.

Lunati

Lunati offers both upgraded OEM rockers and Voodoo aluminum rockers for LS applications. The GM LS Retro Fit rockers are GM cast steel, upgraded with captured bearing heavy-duty trunnions, available for both LS1/LS2 and LS3/L92/LS7 sets with offset intakes. Available ratios are 1.7:1 and 1.8:1 for all except LS7, which is available only in 1.8:1. The extruded aluminum, CNC-machined Voodoo roller rockers are designed with high-lift/spring clearance in mind.

Scorpion Racing Products

Scorpion aluminum roller rockers are available in both 1.7:1 and 1.8:1 ratios, in either pedestal or U-channel radiused pocket mount. According to Scorpion, standard pushrod length applies and there is no need for valve cover spacers for added clearance. They are anodized in the company's signature blue color.

T&D Machine Products

T&D offers shaft-mounted rocker assemblies for LS1/LS2, LS3/L92, LS7, Trick Flow GenX LS1/LS2, and World Products Warhawk LS7 applications. The LS3/L92 set features a .215-inch intake offset. The shaft-mounted set specifically for Trick Flow heads features a .130-inch offset. The LS7 and Warhawk LS7 versions feature an intake offset of .350 inch. T&D also offers several options for its rocker arm sets, including a .040-inch spring oiling hole to deliver added oil to the valve springs, a steel-body option, shot-peening, a lightweight option that carves away unnecessary mass, needle bearing tips, and an aluminum mounting stand option for reduced static weight.

Yella Terra

This firm offers its Ultra Lite aluminum roller rockers that are shaft-coupled as intake and exhaust pairs. LS1/LS2 versions are offered in 1.7:1 or 1.8:1 ratios. LS3/L92 sets feature a 1.7:1 ratio with offset intakes, and LS7 versions are 1.8:1 ratio with offset intakes.

Oiling Systems

Standard wet sump oiling is featured on all LS engine platforms except LS7 and LS9 engines that were equipped with dry sump systems. The wet sump system's oil pan contains the engine's oil supply. A crankshaft-driven oil pump is mounted up front at the base of the crankshaft snout behind the timing cover. Instead of an oil pump that is camshaft driven at half engine speed, the LS pump is driven at full engine speed by the crankshaft. As the crankshaft rotates, the oil pump pulls oil from the oil pan sump through a tubular pickup tube assembly and distributes oil throughout the engine's oil passages, flowing through the oil filter before entering the rest of the engine.

Oil pressure depends on oil viscosity, engine wear, and temperature. The lower the viscosity, the lower the pressure at idle. The higher the viscosity, the higher the pressure at idle. With increased bearing wear, oil pressure drops. As temperatures rise, the oil thins out and pressure becomes lower at idle. As a rule of thumb, the engine should produce about 10 psi of oil pressure for every 1,000 rpm of engine speed. Typical LS oil pressures will be around 15 to 20 psi at idle, about 40 psi at cruise, and about 60 psi at wide open throttle (WOT). Again, variables such as viscosity, engine wear, engine speed, and temperature will affect pressure. Oil pressure at idle should not be below about 15 psi. However, 7 psi provides the bare minimum required to maintain lubrication.

Wet sump oil pumps feature a gerotor design wherein an eccentric gear rotates inside the pump housing to produce pressure. To access the oil pump, removal of the crankshaft pulley and front cover is required. Removing the timing chain set is not needed. The pump is secured to the block with four 8-mm x 1.25 x 20-mm screws, torqued at 18 ft-lbs. No gasket is required between the pump and block.

The LS oil pump in both wet sump and dry sump applications is directly driven at full engine speed by the crank snout. Servicing the pump requires removal of the water pump, crank pulley/damper, and front cover.

The pickup tube secures to the oil pump with a single 6-mm x 1.00 bolt, torqued at 106 in-lb. A single O-ring seal installs between the pickup and pump. Note that if you use an aftermarket oil pan, it may include a pickup tube that is dedicated to that pan. The pickup tube may include a thicker O-ring due to a design change of the aftermarket pickup. Pay attention to the instructions if you use an aftermarket pickup.

An oil galley on the left (driver) side of the block runs from the front

A wet sump LS pump connects to its pickup tube. The tube enters the pump, sealed with a single O-ring. The pickup tube bracket secures to the pump with a single 6-mm bolt.

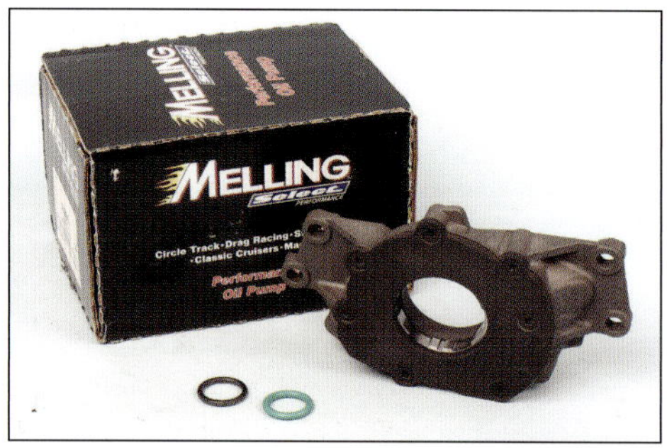

Direct replacement wet sump oil pumps are available from manufacturers such as Melling Select and offer greater pressure and/or volume choices.

Shown here is a wet sump pickup tube mounted. A stud at the number-3 main cap provides additional support for this lengthy tube.

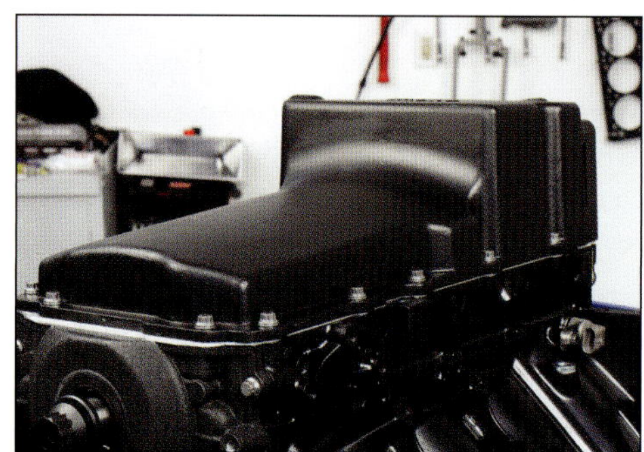
OEM and aftermarket wet sump oil pans are secured with a series of 8-mm screws. The cast-aluminum oil pans are very rigid and actually serve as structural components, providing support to the LS Y-block pan rail contact areas.

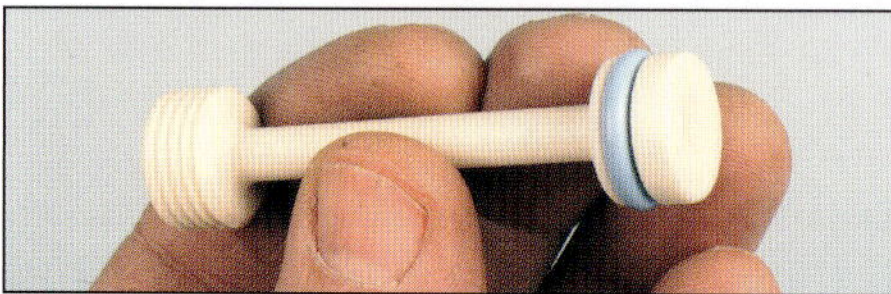
LS engines feature an oil diverter plug, often referred to as a "barbell." This is inserted into the rear of the block. The rearward O-ring end seals oil from exiting the block, while the stem section allows flow to the oil filter boss. This plug is available as GM PN 12573460.

Insert the barbell plug until the plug face is flush with the block. Make sure that this plug is installed prior to installing the rear engine cover.

At the front left side of the block face, a small 16-mm expansion plug seals off the same oil galley where the rear oil diverter barbell plug is installed. Add a small amount of sealer to the outside of the plug cup prior to installing. This plug is available as GM PN 9427693.

Install the front plug until it's flush with the block face.

of the block to the rear. The front passage is located immediately next to the oil pump to the right of the pump when viewing the engine from the front. This front hole is plugged with a small expansion plug. At the rear, on the left side of the block, is the opposite end of this galley. A plastic O-ring plug, often called a "barbell" because of its shape, inserts into this galley opening. This plug serves two purposes: to seal oil at the rear of the block and to allow oil to be diverted to the oil filter boss. The barbell plug is installed with the bare end inserted first, toward the front of the engine, with the O-ring end at the rear of the block. Install until the rear of the plug is flush with the block surface.

During a rebuild, oil passages must be open to aid in cleaning the block. Try removing the rear plastic plug first. If it's stuck, try picking at it and prying it out with a small flathead screwdriver. Don't worry about ruining the plug because new ones are available for about $10 or less. To remove the front expansion plug, run a 1/4-inch- or 5/16-inch-diameter rod, about 2 feet long, from the rear of the galley forward and tap the plug out using a hammer against the rod.

A new rear main seal includes a white nylon guide ring. This guide prevents the seal lips from folding rearward during installation onto the crank flange.

With the new seal installed to the rear cover, the plastic seal guide must be in place.

As the rear cover is positioned to the block, the plastic seal guide slips off and may be discarded. Do not apply lubricant to the seal or crank flange prior to installation. The seal should be installed dry. Oiling the seal can cause the lips to fold back, even when using the seal guide.

Once the rear cover is installed, closely inspect the seal to verify that the lips have not folded to the rear. If the lip has bent rearward, the seal will leak.

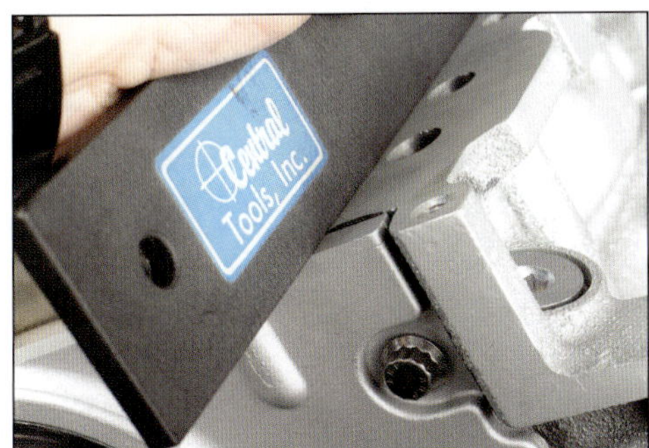

Before tightening the rear engine cover fully, use a straightedge across the bottom of the cover and the block's pan rails to verify that the adjacent surfaces are inline. This helps obtain a superior oil pan gasket seal. The same holds true for the engine's front cover.

Rear Main Seal

The crankshaft rear main seal is installed to the rear engine cover. If you purchase a new rear cover, a new seal is already installed. If you're performing a rebuild, always replace this seal. When installing the rear cover to the engine block, you must use a disposable white nylon seal guide. This is included with the new rear seal. Do not apply lubricant to the rear seal lips. The rear seal should be installed to the crank rear flange dry. Insert the nylon seal guide into the seal opening. Install a new rear cover gasket to the cover. Install the rear engine cover squarely, avoiding an angled approach. Push the rear cover into place. This will cause the nylon seal guide to pop off.

Once the rear cover is in place, finger-install all rear cover 8-mm bolts but do not fully tighten yet. Using a metal straightedge, verify that the bottom of the rear cover is flush and inline with the block's oil pan rails, then final tighten all cover bolts to 18 ft-lbs.

High-Pressure/Volume Upgrade

The engine's horsepower level alone is not the deciding factor in choosing wet or dry sump oiling. Rather, the oil delivery system selection is more importantly based on the vehicle's ground clearance and how the vehicle will be operated. If the vehicle has minimal ground clearance, a wet sump system offers an increase in clearance with its relatively shallow height profile of the pan. The pan serves only as a lower crankcase cover, with the oil supply stored in a remote reservoir.

A dry sump system makes more sense if the engine is subjected to severe and/or sustained angles of operation and either lateral or longitudinal g-forces, such as in drag racing, road racing, or oval track racing, because it eliminates the concern of the wet sump pickup being uncovered during severe acceleration or cornering forces.

While the level of horsepower alone doesn't dictate the need to use a dry sump system, as the power level increases, chances are the vehicle will be used in a more severe manner with regard to acceleration and cornering. For mild to moderate street operation only, either system should be acceptable as long as the oil pan isn't contacting anything.

Aftermarket performance wet sump pumps are available from

manufacturers, such as Melling Select and others, for 4.8L, 5.3L, 5.7L, 6.0L LS2; LQ9; and LS3 6.2L applications. As examples, Melling's 10295 pump provides a 10-percent increase over stock pressure while maintaining stock volume; pump 10296 provides an 18-percent increase in volume while retaining stock pressure. Both pumps include stock and high-pressure relief springs to give you the option of either pressure level. As another example, Lingenfelter Engineering's L200315297 oil pump offers about a 30-percent increase in volume and includes both a stock and high-pressure spring.

Stock pump or direct aftermarket replacements are fine for most applications. However, as engine wear increases, or when a performance engine is built with looser oil clearances, a higher volume of oil is needed to maintain adequate pressure. The use of thicker-viscosity oils and/or colder temperatures tends to generate higher pressure, but not greater volume. As noted earlier, while pressure is important to the tune of about 10 psi per 1,000 rpm, as long as you have enough pressure to provide lubrication, volume is the greater priority. It may be interesting to note that only about 7 psi is needed to provide enough oil film support to "float" the crankshaft on its bearings, although a minimum of about 15 psi is recommended at idle for operating efficiency.

Dry Sump

As mentioned earlier, a dry sump system is by far the preferred approach if the engine will experience severe acceleration or cornering loads. The benefits of a dry sump system include constant oil delivery regardless of lateral or longitudinal forces. A dry sump system also improves scavenging, reducing parasitic oil cling to the rotating assembly, which reduces drag and crankcase pressure. Oil pressure can also be adjusted via changes to the pump gearing. And, because the oil supply is stored in a remote reservoir, a greater volume of oil can be maintained, which helps to cool the oil before it returns to the engine.

LS7 and LS9 oil pumps mount in the traditional LS location at the crank snout. These pumps feature a dual role, delivering pressurized oil to the engine and scavenging oil from the engine and the remote oil tank. Direct replacement high-performance pumps for LS7/LS9 applications are available for boosting pump performance. Shown here is Katech's KAT-A5069, which mounts in the stock location but provides a 30-percent increase in scavenging capacity and 20-percent greater pressure output capacity. It is CNC-ported for superior flow. (Photo Courtesy Katech)

Dry sump systems, depending on the number of scavenging stages in the pump, offer greater flexibility in terms of creating vacuum in specific areas of the engine, as in the lifter valley. If we were to arbitrarily choose a horsepower range wherein a dry sump system might be preferred, this would likely be in the 600-plus-hp range, again anticipating how the vehicle might be operated. That does not mean that an engine that produces 600 or more horsepower absolutely requires a dry sump system. Vehicle operation is a critical factor. A dry sump system is more costly and more complex as opposed to a wet sump system, but if you have the budget and are mindful of the system maintenance, going with a dry sump system is always beneficial.

LS7 7.0L engines and LS9 6.2L engines were fitted with dry sump oiling systems. However, compared to a traditional dry sump system that uses an external belt-driven pressure and scavenge pump, the LS7/LS9 features somewhat of a "budget" dry sump system. It uses a crank-snout-mounted pump (similar to all LS wet sump system pumps), but the pump is thicker front to rear to accommodate the additional gear drive; hence the longer crankshaft snout needed to provide the extended drive surface for the larger pump. The factory pump is actually two pumps in one, and it both delivers oil to the engine and scavenges oil from the remote oil tank.

A traditional competition dry sump system consists of an external belt-driven oil pump, a lower crankcase cover (in lieu of an oil pan that features a sump), and a remote oil reservoir. The pump draws oil from the remote reservoir and pumps pressurized oil into the engine, with oil then being scavenged and returned to the remote reservoir tank. The bottom cover simply serves as a protective cover to seal the crankcase.

As mentioned earlier, dry sump systems offer several advantages. Ground clearance is one benefit. With the elimination of the oil pan's

This LS9-equipped Corvette features a reservoir oil tank in the right rear corner of the engine bay for easy access. Note the factory designation for the use of 5W30 oil, and a dipstick for checking oil level in the tank.

sump, the engine may be mounted lower in the chassis and/or the vehicle can better accommodate a lowered stance. Also, the pump directly injects oil into the engine at key locations, producing more efficient oil delivery. Oil is not stored in a pan sump, so oil is constantly pressure-injected into the oil galleys; the concern for oil starvation is eliminated during severe lateral and longitudinal forces under braking, turning, and acceleration. We're not dealing with a submersed pump pickup in an oil pan wherein the oil supply could potentially move away from and expose a wet sump pump's pickup. In addition to eliminating those concerns, the dry sump pump also pulls vacuum from the crankcase, so windage and crankcase pressure are reduced because oil isn't sloshing around in the pan. This reduces blowby, allowing you to use lower-tension piston rings, further reducing parasitic friction, with potential horsepower gain.

Dry sump pumps are offered in "stages," referring to the number of scavenge plumbing circuits. For example, a two-stage pump pulls, or scavenges, from the engine pan and pulls delivery oil from the reservoir tank to the pump, with a single pressure line delivering oil from the pump to the engine. A five-stage pump, as another example, might feature scavenging from the oil tank, from the engine's lifter valley, and from three locations in the engine pan. Oil is then sent from the pump to both the engine and a remote tank. The number of stages is based on the volume of oil you wish to pull and from which locations. The larger the tank capacity and/or the greater the oil viscosity, the more stages that are required.

A dry sump system is not inexpensive, and it takes up a fair amount of underhood real estate. Cost estimates include a dry sump pan at around $600 to $800, the pump at around $1,100 to $1,400, a remote oil filter at around $100 to $160, a remote oil tank at around $400, a vent for the tank at around $60 to $90, and plumbing at around $300 to $400. Add to this a hub extension if your crank has the traditional short snout and the entire system can run in the area of $3,000 to $3,800. Some complete aftermarket dry sump systems, depending on manufacturer, can run upward of $5,000. However, if you already have an LS7 or LS9 that features a factory dry sump system, you can upgrade the pump with a high-performance direct replacement unit for around $300 to $400.

Shown here is Katech's Ultra Track Attack LS7 engine package, an example of a highly modified LS7 featuring an external dry sump pump. Note the belt-driven pump at the lower right (passenger's side) of the block. (Photo Courtesy Katech)

Converting to a dry sump system or planning to perform a build with a dry sump system is relatively straightforward, thanks to the many aftermarket offerings. Citing AVIAID as but one example, it offers four different dry sump levels specifically designed as bolt-ons for LS engines, including a street system that uses the OEM pump for pressure and an external pump for scavenging, a Club Racer system that mounts the external pump to the A/C bracket, a three-stage scavenge/pressure pump that mounts to the block, and a competition system with a four- or five-stage pump.

Aftermarket dry sump systems and components are offered by a number of sources, such as AVIAID, Jones Racing, Barnes, Peterson Fluid Systems, and a host of others. Refer to the aftermarket source listing for a complete list and contact information. Just be aware that in many cases, you'll need to use a longer-snout crankshaft that's part numbered for an LS7 application. These are available in a variety of strokes to suit your needs.

Oil Pressure Sensor

Most LS engines feature the oil pressure sensor at the top rear left of the block, behind the valley cover. The hole features 16 x 1.5–mm threads. If you plan to use an aftermarket oil pressure gauge, this sender will not be compatible with the new gauge. Aftermarket gauge suppliers offer an adapter that features 16 x 1.5–mm male threads and a 1/8-inch NPT female thread to allow easy adaptation of the aftermarket gauge. An option is to use an old factory sender by cutting off the plastic top and tapping a 1/8-inch NPT hole in the center.

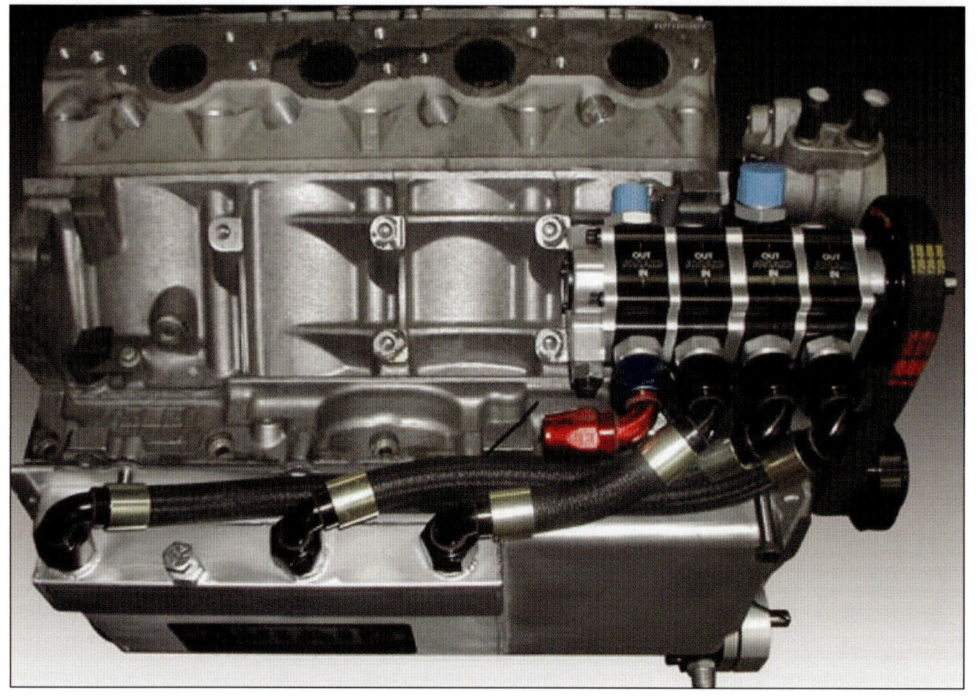

Note the three scavenging connections to the engine pan on this AVIAID dry sump pump mounted to an LS engine. (Photo Courtesy AVIAID)

AVIAID offers a basic external two-stage dry sump pump kit for LS applications for LS owners who wish to upgrade from wet sump to dry sump at a lower cost. (Photo Courtesy AVIAID)

The example shown here is an LQ9 iron block with the oil pressure sender installed. Note the bore for the early–Gen III rear-mounted cam position sensor.

Original oil pressure sensor locations on most LS engines are at the rear left of the valley plate, although LS7 and LS9 dry sump locations may be at the side of the engine lower pan at the filter boss. The common GM PN is 12616646, featuring a 16 x 1.5–mm thread.

If you're performing an engine swap and/or running a carburetor and wish to install an aftermarket oil pressure sender, you can use an original factory sender by chopping off the upper plastic connection body and tapping a 1/8-inch NPT hole. In this photo, a 1/8-inch NPT plug was temporarily installed to seal the new hole.

EFI and Intake Manifolds

The LS engine platform is versatile; as such, you have your choice of carburetion, central fuel injection, or multi-port injection. It all depends on the application and performance goals. The advantages of running a carburetor include lower cost and a more simplified system. Central injection, also called throttle body injection (TBI), involves running a carb-style intake manifold with a throttle body that features built-in injectors. Multi-port injection provides a "factory" approach, utilizing an air throttle body with an injector dedicated per cylinder, injecting fuel into each intake runner.

Running a carbureted LS engine is extremely simple and requires no modifications. The fuel system requires no electronic management. Simply bolt on a carb-style intake manifold and carburetor. For ignition, you'll need an ignition controller, a set of eight coil packs, and the harness to connect the controller and coils. MSD, for example, offers an 6LS ignition controller for either 24-tooth reluctors or a 6LS2 controller for 58-tooth crankshaft reluctors. The MSD harness connects the controller to the coils, the cam position sensor, and the crank position sensor. No programming is needed. The ignition controller includes a set of preprogrammed plug-in "chips," each with its own ignition curve. Or, you can create a custom curve using the included CD and a laptop or desktop computer. For those who prefer the appearance, simplicity, and lower cost of a carb setup, this is the easiest and least expensive approach. ECU is not needed; only the small ignition controller is required. MSD also recently introduced an LS ignition controller that accommodates either reluctor wheel tooth count.

For those who prefer fuel injection, the next level involves TBI, which is often referred to as a "wet"-style throttle body. Using a carb-style intake manifold, an injection-equipped throttle body mounts like a carburetor. This wet-style throttle body

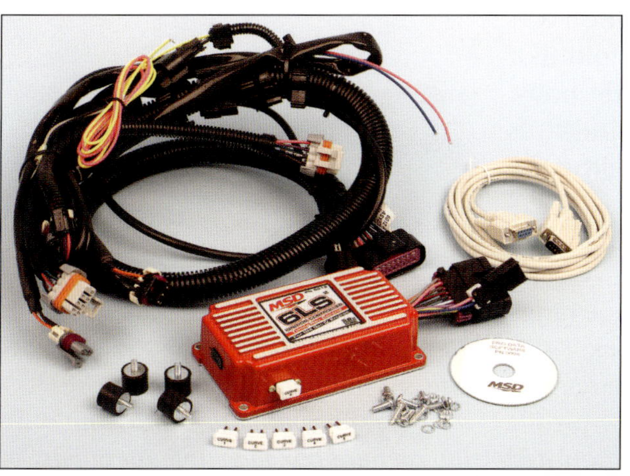

The MSD LS ignition controller kit includes everything needed for installation except of a pair of coil harnesses, which can be purchased from any Chevy dealer (or you can use the coil harness that originally came with your engine). Three versions of controller are available: one for a 24-tooth reluctor, one for a 58-tooth reluctor, or a new version that will work with either reluctor tooth count.

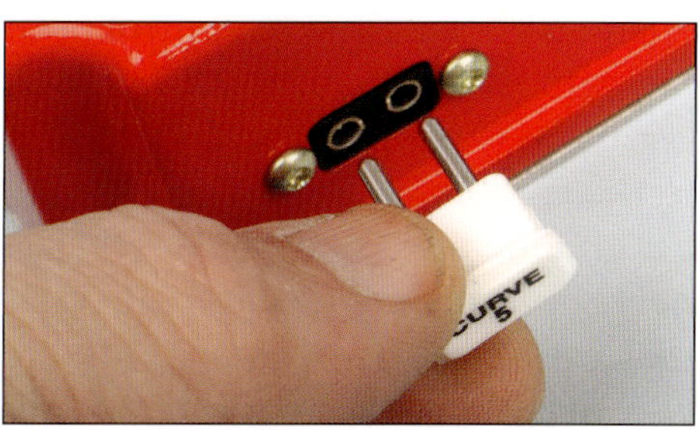

Changing the ignition curve on the MSD controller is as simple as plugging in a different chip. Each chip is numbered. The instruction manual provides the ignition curve that each chip offers.

Be aware that depending on the aftermarket kit you purchase, you may need to also obtain a pair of coil pack harnesses. These are readily available from a GM dealer.

handles both air and fuel in one package, distributing the mixture from the central location above the manifold runners. The throttle body features built-in fuel injectors and somewhat resembles a 4-barrel carb in appearance. A built-in throttle position sensor is featured as well. A kit includes the appropriate wiring harness and a controller, which can be programmed or, depending on the make and model, run as-is, with a self-learning program that adjusts fuel and spark as you drive. Toss in a set of eight coil packs, and you're ready to go.

Another option is to use a carb-style intake manifold, an air throttle body, and eight individual fuel injectors that are placed in bungs at the base of each intake runner. This provides a carb-style air intake coupled with direct-port fuel injection. Again, since we need to control both fuel and spark, a controller and wiring harness are required in addition to the coil packs. These systems may be obtained as individual components or as a complete kit. Multi-port fuel injection intake manifolds are also available for mounting the air throttle body up front, referred to as "longitudinal" intake manifolds.

The final option is a factory-style crossover intake manifold that resembles the OEM system in terms of overall appearance and design. Either a factory or aftermarket ECU is required to control both fuel and spark. While respectable horsepower and reliability may be obtained with any of these approaches, fuel injection offers a more precise approach that is less susceptible to temperature, humidity, and altitude issues, but those who are adept at carburetor tuning may feel more comfortable with a carb setup. As an example, my shop has built a number of carb-equipped and TBI-equipped LS engines, producing from 625 to more than 670 hp. Your choice will depend on budget, preferred appearance, and application. Manufacturers such as Holley, Edelbrock, and others offer a wide range of fuel and air delivery components and systems for Gen III and IV LS engines.

Intake Manifolds

For those who wish to run either carburetion or a central fuel injection setup, a host of intake manifolds are available through the performance aftermarket. Holley, for example, offers single-plane, dual-plane, and hi-ram intake manifolds that accept a single or dual 4-barrel carburetor setup or a central injection throttle body such as those offered by MSD, FAST, and others. Another option is to run a carb-style intake manifold that accepts an air intake throttle body, along with per-cylinder fuel injectors via injector bungs machined into the manifold runners.

Multi-port fuel injection manifolds of the longitudinal style (similar to OEM) are available in a wide range of designs by Holley, Edelbrock, FAST, and others. These involve a large plenum with either high- or low-profile designs, accepting a front-mounted throttle body. This style is ideal for those who want to upgrade their existing factory manifolds.

Dart Machinery, a long-time major player in the development of high-performance and racing LS blocks and cylinder heads, now offers two intake manifolds for the LS platform for those wishing to run either carburetors or central throttle bodies. A 10-degree Box Ram intake features tall runners and a billet pent-roof top plate. Another version features a single 4150-style carb or throttle body flange.

Our Holley single-plane intake manifold PN 300-132BK features a black ceramic coating and cathedral ports (LS1/LS6/LS2 style) and is designed for an optimum operating range of 2,500 to 7,000 rpm. Ports measure 2.66 inches high x .92 inch wide. The 4150-style square bore flange accepts up to 1.75-inch-diameter throttle bores, and the height from the valley plate to the carb flange is 4.95 inches.

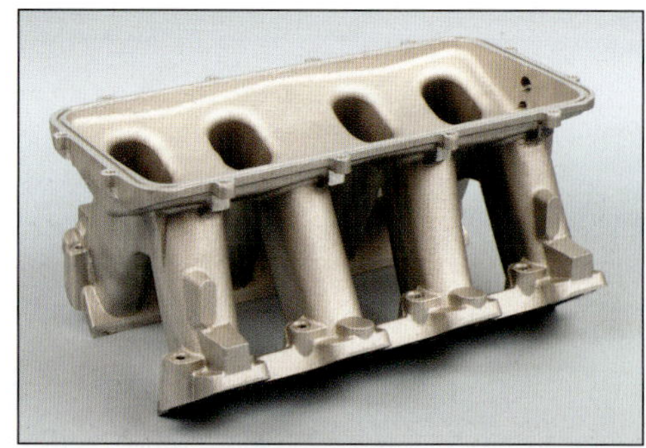

The Holley Hi-Ram intake PN 300-226 offers a modular design, allowing a single- or dual-carb setup, with a top plenum box available for either approach.

Either manifold is available with or without injector bungs. The extra-tall/long runners provide higher horsepower and an extended RPM range.

Throttle Body Injection

As mentioned earlier, throttle body injection (TBI) involves a throttle body with built-in fuel injectors in one neat package. Examples include

Dual-carburetor plenum PN 300-216 is shown here. The plenum bolts to the manifold and is sealed with an O-ring strip.

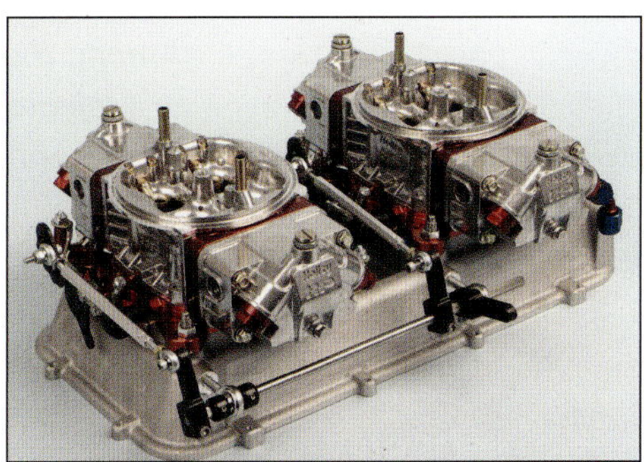

A pair of Holley 600-cfm Ultra carbs are mounted "sideways" on the plenum to facilitate a throttle linkage system. This induction setup on a 408-ci LS engine equipped with Trick Flow heads pulled 670.5 hp at 6,500 rpm on the engine dyno and 581.7 ft-lbs of torque at 4,900 rpm. Perhaps not a practical street setup, but it sure looks cool and loves high RPM.

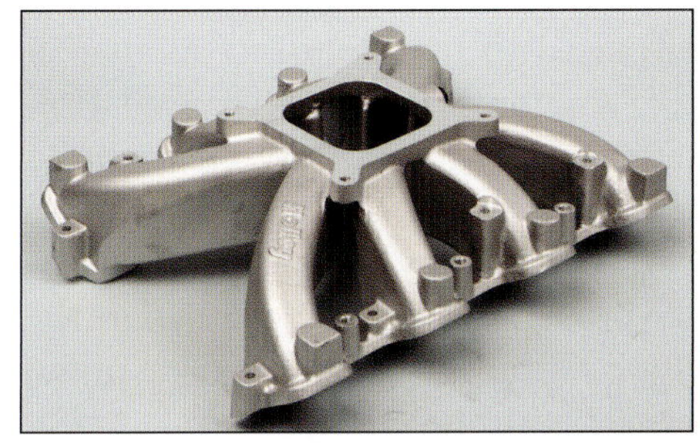

In addition to square-bore manifolds for LS cathedral-port heads, Holley also offers a single-plane intake with LS3-style rectangular ports as well as with LS7 ports.

During one particular build, we fitted LS3-style Trick Flow heads and a Holley LS3-style manifold along with a Holley 850-cfm carb. Interestingly, we tried a 1-inch carb spacer and picked up another 10 hp on the dyno.

Holley's Sniper longitudinal EFI intakes are fabricated sheet-aluminum units as opposed to cast. They are available for various throttle body sizes, in cathedral or rectangular ports, in natural, polished, or black. (Photo Courtesy Holley)

Holley's EFI Hi-Ram longitudinal intakes are available to accept 92- to 105-mm throttle bodies. These intakes are available for either cathedral-port (LS1/LS6/LS2) or rectangular-port LS3/L92-style heads, with finishes in natural or ceramic black. Injector bungs are featured, to be fed with a billet fuel rail at each bank. (Photo Courtesy Holley)

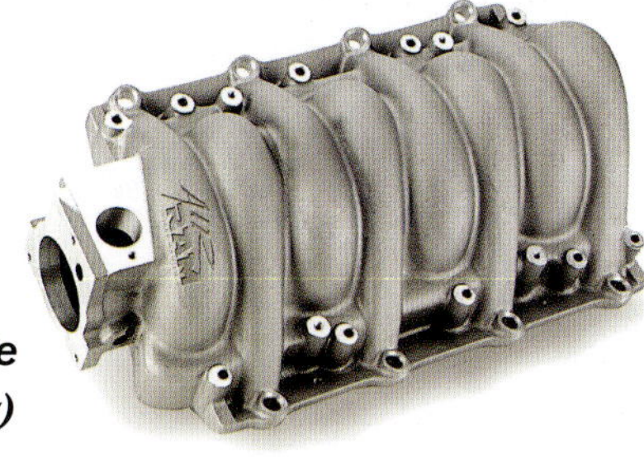

Weiand's cross-plenum EFI intake is a cast-aluminum unit, available for cathedral-port applications. (Photo Courtesy Holley)

Among the factory and aftermarket multi-port injector manifolds available, Edelbrock now offers a very cool Cross-Ram intake specifically for LS3-style heads. This features a relatively low profile and long runner design that accepts two 90-mm throttle bodies, making it a great choice for those who intend to run twin turbochargers. These are available with the right and left plenums in either red (PN 7141) or black (71413). (Photo Courtesy Edelbrock)

For the dual-carb setup, a pair of 600-cfm Holley Ultra carbs, PN 80801-RD, worked well on a 408-ci LS build along with a tunnel ram manifold.

Fuel Air Spark Technology (FAST) offers an array of multi-port injection manifolds, constructed of high-strength polymer in a module design of an upper plenum and runner bottom section, allowing porting of individual runners. Selections include the LSX, designed to accept a 92-mm throttle body for cathedral-port heads, and the LSXR, designed for a 102-mm throttle body, with applications including LS1/LS6/LS2/LS3 and LS7 heads. Also offered is the LSXRT, for use with a 102-mm throttle body, for cathedral-port heads only. Previous intake manifolds featured a silver/aluminum color, while recent models have been released in an all-black finish. Depending on the model, increases of 25 to 30 hp are reported compared to stock manifolds. (Photo Courtesy FAST)

Holley's Terminator EFI and new Sniper EFI, FAST's EZ-EFI, Edelbrock's E-Street and Pro-Flow 3, and MSD's Atomic EFI. For those who wish to obtain a simple-to-install electronic fuel injection atop a 4-barrel intake manifold, these systems are the perfect solution. The all-in-one air/fuel unit mounts just like a carburetor, requiring minimal wiring. Fuel injectors, throttle position sensor, and ECU are built into the unit. Most are rated to handle up to about 650 hp. As one example, I've used Holley's EFI Terminator on an LS build that dyno'd at 657 hp. Setup was surprisingly easy and produced a better power and torque curve than the dyno shop's dyno-proven "known good" custom carb.

Aftermarket performance wet-style throttle bodies are available in a range of cfm ratings, sometimes in the 900-cfm-or-greater range. Don't be misled by comparing these ratings to those of carburetors. While a carburetor moves both fuel and air and must be sized according to engine displacement and horsepower, an EFI throttle body will simply make the volume of air that is needed at a specific time.

More recent introductions include Wilson Manifolds' new LS

A FAST manifold and fuel rails installed in a Z06 Corvette.

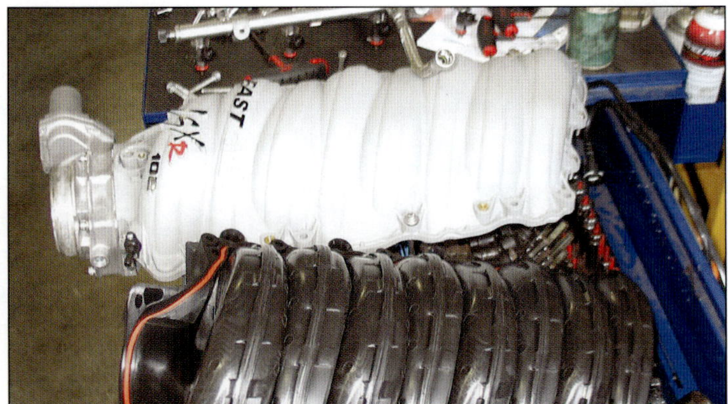

The FAST composite manifolds are modular. With the top plenum removed, individual runners are accessible for custom porting.

Just because you're dealing with an LS engine doesn't mean you're restricted to using fuel injection or ugly factory valve covers. You can easily achieve a radical old-school look by using your choices of carburetion setup and small-block Chevy valve covers thanks to available valve cover adapters. This Hi-Ram Holley intake manifold outfitted with dual 600-cfm Holley carbs and tall small-block Chevy valve covers is a good example. The tall velocity stacks added a nice, radical touch. This example might be appealing for street rod owners who prefer appearance over practicality, especially for those who view "too much" as "not enough."

Edelbrock's Pro-Flo XT LS injection system features a large plenum and tapered runners for a broad torque band. This setup reportedly provides an increase of 30 hp over a stock LS6 manifold. Applications for this manifold include LS1/LS6 and LS2. The manifold is offered in either satin or black. The management system includes everything needed for installation. (Photo Courtesy Edelbrock)

This LS7-equipped drag car built by Hutter Engineering of Chardon, Ohio, is a good example of a maximum-output naturally aspirated setup. Note the custom Wilson fabricated intake manifold and dual throttle bodies fed via separate carbon fiber intake tubes.

If you're running a carb setup on an LS engine, consider adding a breather directly to the front of the valley cover. This allows more efficient reduction of crankcase pressure as opposed to breathers on the valve cover(s) only. This involves boring a hole in the valley cover to allow welding an aluminum breather tube, then attaching a breather element. Position it as close to the front intake manifold runners and carb as possible while still providing enough clearance to service the breather element.

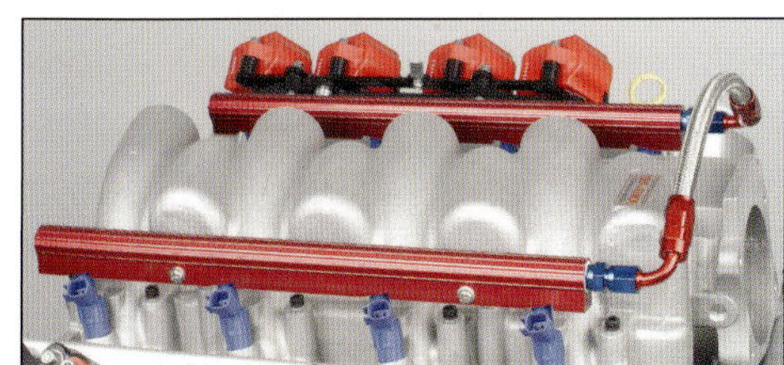

Billet-aluminum fuel injector rails commonly allow plumbing crossover hoses in –6 AN or –8 AN size. The –6 is equivalent to a 3/8-inch inside diameter; –8 is equivalent to a 1/2-inch inside diameter. Rails are also available in a variety of finishes, including polished, black, red, or blue.

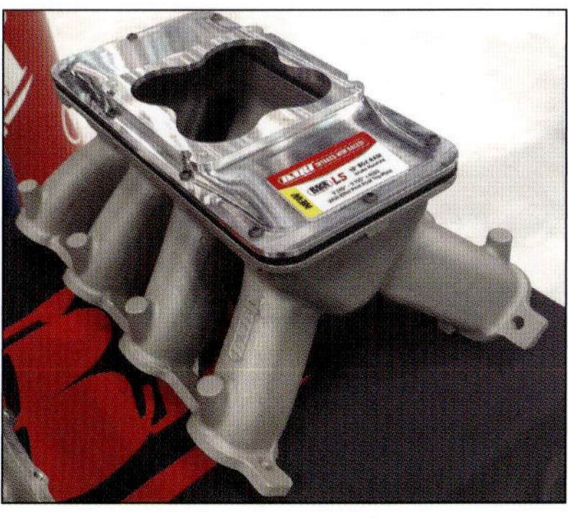

Dart Machinery has also developed a line of high-flowing tall-runner intake manifolds. This version is its 10-degree Box Ram manifold equipped with a billet pent-roof top plate. It is available with or without injector bungs.

Dart also offers a 4-barrel long-runner manifold for a single 4150-style carb or throttle body, also available with or without injector bungs. The tall, long runner design definitely will boost the RPM range and will increase power at higher engine speeds.

applications for EFI setups, including one with a flat billet top plate that allows mounting a pair of 4-barrel air throttle bodies and one with a low-profile system that accommodates a single front-mount throttle body and includes extra nitrous injection ports. All components are CNC-machined from billet-aluminum stock. Wilson is well established for providing induction systems for professional racing applications that also apply to high-performance street use.

A sleek new low-profile injection manifold from Plazmaman (an Australian firm) features 100-percent billet-machined construction made of 6061 aluminum, available in port sizes to accommodate LS1, LS3, and LS7 ports. Nitrous compatibility is available with a 16-injector version and built-in burst panels. Finishes are available in clear or black anodized surfaces. Yet another offering is from Magnatron, featuring a CNC-machined lower intake and curved upper plenum, designed to accommodate any size throttle body with available adapters.

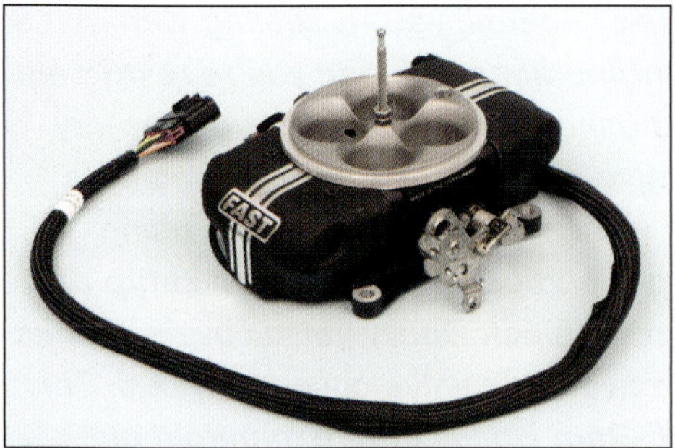

The FAST EFI throttle body provides the appearance of a carburetor, with multiple injectors hidden behind faux fuel bowl covers. The throttle body features two harnesses, one that connects to the EFI controller and one that connects to the water temperature sensor on the left cylinder head.

The FAST EZ-EFI 2.0 kit includes everything required for fuel management. Our kit even included an inline high-pressure fuel pump and adjustable regulator. For LS applications, the FAST EZ LS ignition controller, which connects to the EFI main harness, is also required.

An example of the FAST EFI throttle body injection mounted to a Holley single-plane intake manifold. The outward appearance satisfies those who prefer a carburetor appearance, combined with the higher fuel delivery precision offered by EFI. Utilizing a TBI unit allows installation of a 4-barrel-style air cleaner. The faux "fuel bowl" covers hide the injectors from view.

A pair of fuel injectors is featured on both the primary and secondary sides of the throttle body. Unlike other TBI units that cover the injectors, the injectors are visible and easily accessed for service to mimic the look of a carburetor.

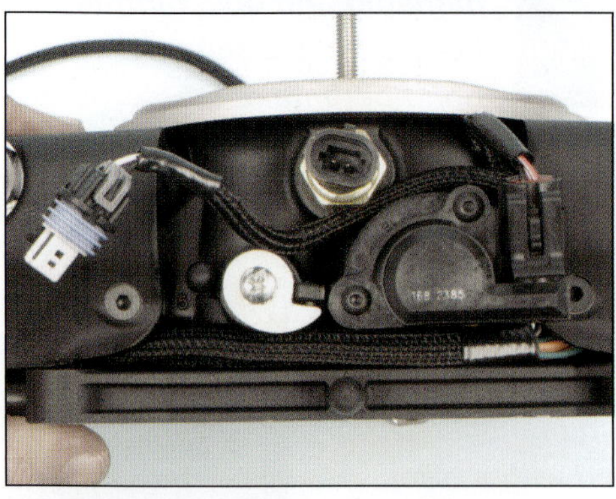

The right side of the FAST throttle body features a built-in throttle position sensor.

The FAST EZ-EFI control module. This can be installed in the vehicle interior.

The FAST EFI throttle body injection system features a handheld control module that can be temporarily connected during setup or mounted in a convenient in-car location, allowing quick and easy adjustments.

The Holley Terminator EFI system includes everything needed for installation. Setup is fairly easy and straightforward, once you take the time to read the setup instructions! The dark-charcoal anodized finish gives a pro appearance.

The Holley Terminator EFI features 950 cfm of potential airflow, with four integrated 65-pound fuel injectors. Don't be intimidated by the large airflow capacity. Only enough airflow will be utilized based on the air/fuel delivery requirements at any given engine speed. The 900-cfm capacity is simply the available airflow volume.

The Terminator EFI is available in Hard-Core Gray (as shown here) and in a fully polished version. The Terminator throttle body unit is compact and as easy to install as a carb, featuring two fuel fittings, one for feed and one for fuel return.

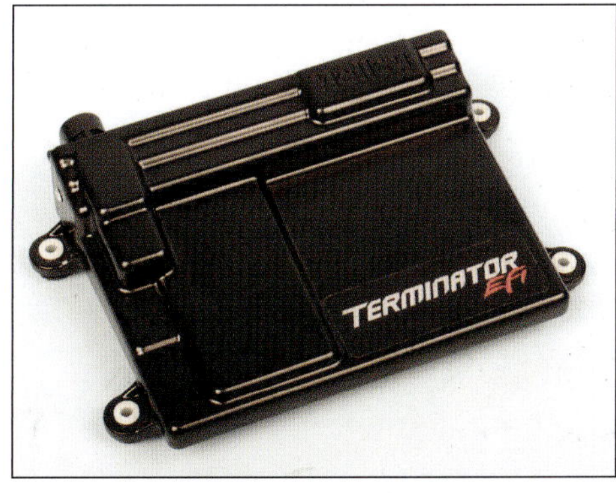

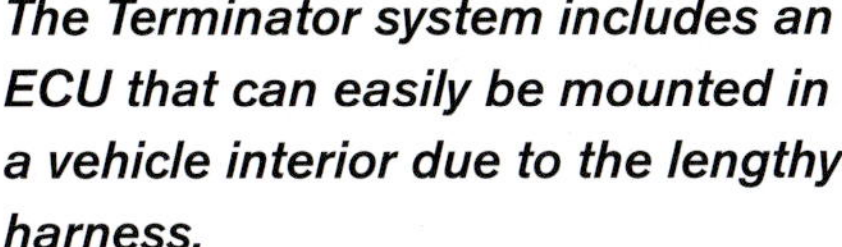

The Terminator system includes an ECU that can easily be mounted in a vehicle interior due to the lengthy harness.

Holley's new Sniper EFI is available in polished, black, or gold. Four injectors, a throttle position sensor, ECU, and fuel pressure regulator are all built into the bolt-on unit. These throttle body injection units are rated as handling up to 650 hp. (Photo Courtesy Holley)

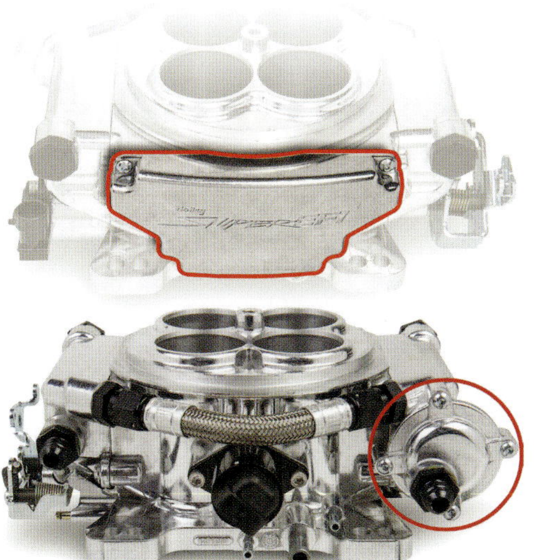

The Holley Sniper EFI features a built-in ECU (top image at left) and a built-in fuel pressure regulator (bottom). (Photo Courtesy Holley)

EFI Alternative

Many performance enthusiasts are familiar with Borla, makers of performance exhaust systems. However, you may not be aware of that firm's entry into the induction market. Borla Induction specializes in fuel injection systems that feature multiple downdraft and sidedraft setups that mimic the appearance of old-school downdraft or sidedraft multiple-carb systems. These assemblies are extremely cool, offering an alternative to more commonplace EFI systems. Complete system offerings include Ford FE, Ford and Chevy small-block, and more, but recently they've introduced setups for the LS platform. They're pricey, but extremely exotic in appearance. The systems combine the vintage look of multiple-carb road race induction with the efficiency of electronic fuel injection. If your budget allows, you need to check this out.

MSD Atomic TBI is designed to support up to 625 hp when used with the specified high output fuel pump. (Photo Courtesy MSD)

Wilson Manifolds, long known for its race-application CNC and welded aluminum intakes, has released two LS manifolds. The example shown here features a CNC-machined EFI intake with a billet machined upper plenum cover that accepts a pair of billet 4150 or 4500 flange throttle bodies.

Wilson also offers an LS EFI intake manifold with central upper plenum that accepts a range of throttle body sizes and includes extra injection bungs to accommodate a nitrous system.

This hard-wired handheld controller plugs into the main harness and allows system setup, calibration, advanced tuning, on-the-fly programming, and system adjustments when located within the driver's reach. The handheld controller may be connected at all times but is not necessary if no additional tuning is desired. You must read the instructions for installation and setup of the EFI system. Don't try winging this. The instructions for setup are lengthy but very easy to follow. Don't be intimidated by the length of the instructions. If you take the time to read, setup will be a breeze. The controller's main menu offers selections for gauges (a graphic view of various sensor readings), monitor (numeric views of sensor readings), wizard (creates base calibration and performs TPS setting), tuning (for adjustment of various parameters), and file (allowing you to load and save numerous calibration settings). The amount of information available on the controller is amazing.

Throttle Body Size

When upgrading to a larger throttle body on a multi-port EFI engine, what size air throttle body is best? It depends on the anticipated engine speed at the engine's peak horsepower. A smaller throttle body increases air velocity, and a larger throttle body slows air velocity. Stock LS1/LS6 engines were equipped with three-bolt 75-mm throttle bodies, while LS2 and LS3 engines featured four-bolt 90-mm throttle bodies. Popular aftermarket throttle body sizes range from 90 mm to 105 mm in size, with 90 mm to 95 mm applicable to modified street and street/strip applications. The 105-mm throttle bodies are designed for maximum airflow in wide open throttle (WOT) applications, such as drag racing.

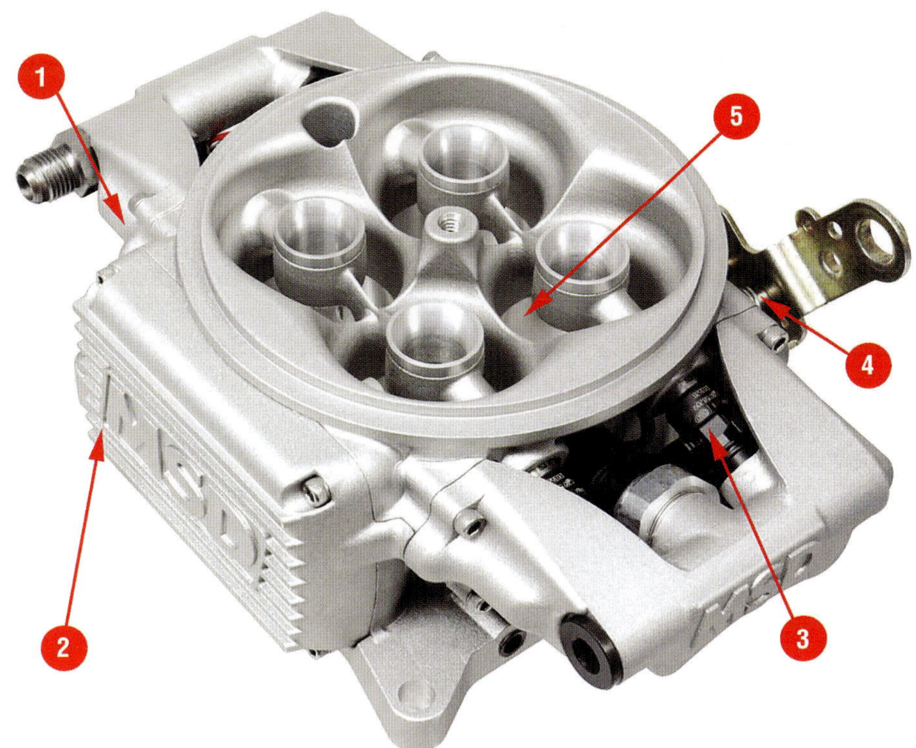

The MSD Atomic EFI features the following:

1. *Internal fuel rail that feeds all four injectors.*
2. *ECU integrated into the throttle body, reducing wiring. TPS (throttle position sensor), MAP (manifold absolute pressure) sensor, IAT (intake air temperature) sensor, and fuel pressure sensors incorporated into the ECU. The only sensors needed for connection are the coolant temperature and wide-band oxygen sensors, which are included in the kit.*
3. *Fuel delivery handled by four 80-pound injectors.*
4. *An automatic, self-calibrating, noncontact sensor TPS, so no setup configuration is needed.*
5. *A throttle body that bolts in place of a standard square bore carburetor and accepts most common throttle and kickdown linkages. (Photo Courtesy MSD)*

An Australian company, Plazmaman, has recently released a series of CNC–billet-machined 6061 aluminum low-profile intakes for the LS platform that accept either OEM or aftermarket throttle bodies. The sleek profile and incredibly detailed exterior finish appear to be easy to clean, and this model is offered in either a silver/clear or black "stealth" anodized finish. It is available with 8 or 16 injector ports (added ports for those who plan to add nitrous injection) for LS1, LS3, and LS7 port applications.

Fuel Injector Size

Follow this handy formula to calculate the optimum injector size for your application:

(Max HP x BSFC) /
(Number of Injectors x Duty Cycle) =
Injector Size in Pounds Per Hour

Brake specific fuel consumption (BSFC) represents the amount of fuel consumed per unit of power produced. BSFC can be found by engine dyno testing, or you can follow these generic guidelines:

- For a high-compression engine of approximately 10.5:1 or higher, running on gasoline, BSFC is about .45 for a stock or moderate build, and about .55 for a highly modified engine.
- For a modified forced-induction engine (supercharged or turbocharged), BSFC is about .55 to .65.

Duty cycle represents the amount of time that a fuel injector is open versus the total time between cylinder firing events. A safe duty cycle for a stock or mild engine is .80, and .85 for modified engines.

As an example, let's say you plan to obtain about 600 hp with a naturally aspirated gasoline engine that has been highly modified. In this example, we'll use a BSFC of .55 and a duty cycle of .85.

(600 hp x .55 BSFC) /
(8 Injectors x .85 Duty Cycle)

So, 330 / 6.8 = 48.5 pounds per hour. As another example involving moderate modifications, where we anticipate 500 hp, we'll use a BSFC of .45 and a duty cycle of .80.

Aftermarket throttle bodies are available in a variety of sizes and offer smooth, radiused air inlets and precision-fit blades. For the appearance-minded, they're also offered in various finishes. (Photo Courtesy Holley)

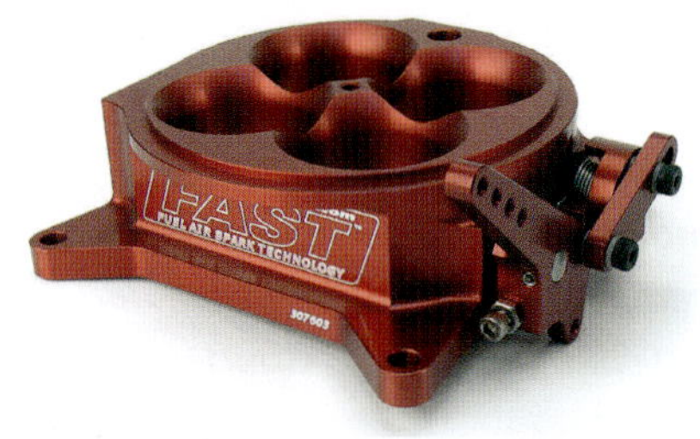

Dry-style throttle bodies are available in machined aluminum in various anodized color finishes.

(500 hp x .45 BSFC) / (8 Injectors x .80 Duty Cycle)
225 / 6.4 = 35.156 pounds per hour

The above examples determine injector flow rating as a starting point. To finely tune the system, an adjustable fuel pressure regulator can be used to modify the flow rating. By noting the fuel injector's static flow rating at a specific fuel pressure, you can calculate a change in injector flow rate by altering pressure. Injector manufacturers will be able to provide an injector's static flow at a specific level of fuel pressure. For example, an injector may be rated at 39 pounds per hour at 40 psi. By increasing fuel pressure, the injector flow rating can be increased.

The formula for determining an increase in flow based on an increase in fuel pressure is as follows:

(Square Root of New Pressure / Old Pressure) x Old Flow Rating

As an example, an injector has a published pressure of 35 psi and a static flow rating of 39 pounds per hour but we want to see how an increase to 42 psi of fuel pressure will affect the flow rating.

42 (New Pressure) / 35 (Old Pressure) = 1.2
The square root of 1.2 = 1.0954451150103321
1.0954451150103321 x 39 (Old Flow Rating) = 42.7 pounds per hour

In this example, we can see that by boosting fuel pressure from 35 psi to 42 psi, we can change the flow rating from 39 pounds per hour to 42.7 pounds per hour.

Note that fuel injectors require higher fuel pressures than do carburetors. While a carburetor requires pressure in the range of 4 to 6 psi, depending on the application, fuel injectors may require 35 to 70 psi. Obtaining the optimum-size injector is a balancing act involving both the flow rating and the fuel pressure.

Fuel injectors are electronically controlled valves that spray fuel into the combustion mix on demand.

Low Impedance versus High Impedance

High-impedance resistance injectors (about 10 to 16 ohms) and low-impedance resistance injectors (about 1.5 to 4 ohms) are available. Which type do you need? If you're running an OEM ECU (computer), generally you need high-impedance injectors. These feature a somewhat slower response time. If you're running a performance aftermarket ECU, you need low-impedance injectors, which react more quickly. If you run low-impedance injectors with an OEM computer, you run the risk of damaging the OEM ECU. *Most* injectors that feature high flow rates are low-impedance type. Low-impedance injectors offer a faster opening response time, generate less heat, and are *generally* preferred for aftermarket performance systems, although this can vary depending on the specific aftermarket system.

Choosing the Correct Injector Size

If you buy a complete system kit, the injectors are likely included. If you need to determine your own injector requirements, you can target the appropriate-size injector based on your engine's anticipated horsepower and the engine's brake specific fuel consumption (BSFC) efficiency number. BSFC represents the amount of fuel that the engine will consume, divided by its power output (how many pounds of fuel will be used per horsepower produced in a hour).

Commonly used fuel injectors include the EV1 and EV6 body styles. The EV1 is the "fat" style, often generically called the "Ford" style. The EV6 features a skinnier profile and a different connector (such as the EV6-style injectors used in newer Fords and GM LS engines). The thinner-profile EV6 also is available in different lengths, so between lower O-ring diameter and overall length differences, you can really get

Generic Rule of Thumb for BSFC

The general rule of thumb for BSFC is:

Naturally aspirated engines	.4 to .5
Engines with nitrous injection	.5 to .6
Forced induction (turbo or supercharger)	.6 to .7

For engines running methanol, double the BSFC number.

(Engine HP x BSFC) / (Number of Cylinders x .8)
= Injector Size

Example: A 500-hp V-8 with high compression

(500 hp x .50) / (8 x .8)
250 / 6.4 = 39 pounds

Theoretically, this application calls for a 39-pound injector.

(This is merely a starting point. Depending on the system and as a result of tuning, a "larger" injector may be required.) ■

confused if you're piecing an MPFI system together without knowing what you're doing. For that very reason, it's best to simply purchase a complete system, where the injectors (in addition to impedance and flow rate) are already matched to fit your intake manifold, fuel rail, and harness setup.

Aftermarket performance systems require a fuel injection controller, or ECU, a throttle position sensor, and an oxygen sensor. Depending on the system, sensor requirements may include a knock sensor, intake air temperature sensor, MAP sensor, and coolant temperature sensor. If the engine was originally equipped with a fuel-injection system, the ignition system will already be controlled by an ECM and will likely incorporate cam and crank position sensors. If the engine was originally equipped with a distributor, EFI systems are available that take this into account, providing fuel injector control. Instead of piecing a system together, it's best to purchase a coordinated system that includes the throttle body along with all necessary control accessories.

Eight-Cylinder Injector Application Chart

Chart courtesy FAST. Note that part numbers shown here refer to specific FAST injector numbers. N/A refers to naturally aspirated, with no forced induction.

Injector Size	Eight-Cylinder Set No.	N/A Peak HP	Supercharged Peak HP	Turbo Peak HP
24 lbs/hour	302408	346	288	276
36 lbs/hour	303608	518	432	415
42 lbs/hour	304208	605	504	484
55 lbs/hour	305508	792	660	634
60 lbs/hour	306008	864	720	691
65 lbs/hour	306508	936	780	749
83 lbs/hour	308308	1,195	996	956
95 lbs/hour	309508	1,368	1,140	1,094
160 lbs/hour	3016008	2,304	1,920	1,843

All ratings shown here are based on a reference constant of 45 psi fuel system pressure.

Remember: The injector itself does not make power. Choose the injector size based on realistic engine output. If in doubt, it's usually best to go with the next larger size.

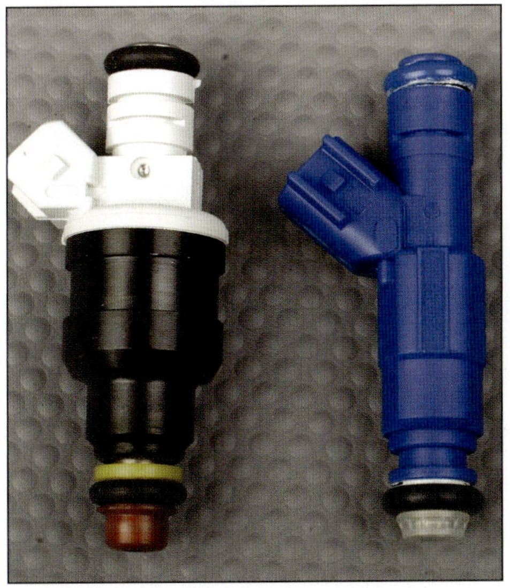

Pictured here are EV1 (left) and EV6 styles. Both styles are available in either low or high impedance.

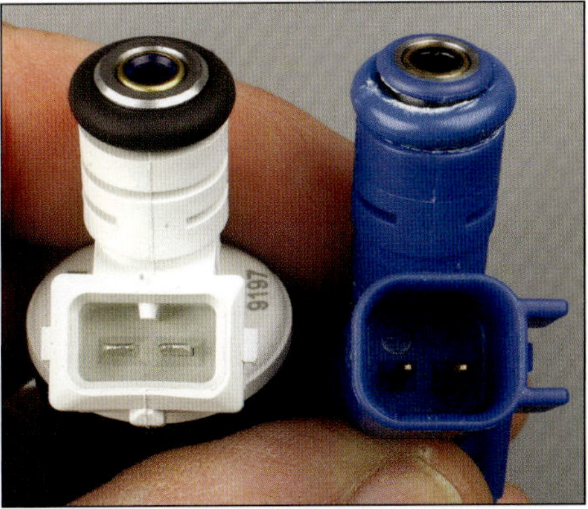

Harness connector styles. At left is the Minitimer style. At right is the USCAR style. Note that the Mimitimer connector features a rectangular housing and flat-blade terminals, while the USCAR style features a square housing and pin-style terminals.

Part II

Two High-Performance LS Builds

The LS engine platform can be built to a wide range of versions, including choice of block and heads, displacement, compression ratio, horsepower and torque peaks, and level of street or competition parameters.

In Part II we list all the components that were selected for two very specific builds, merely to serve as examples. Each example provided includes detailed information to duplicate any of the following engine builds, including block machining, dimensions and clearances, torque specifications, and a complete list of all parts and part numbers used in each build. The following provides a complete road map to build any of the two examples featured.

Dual-Carb LS 408

The first build we discuss features a dual-carb 408-cid, starting with a factory 6.0L LQ9 iron block. This build produced 670.5 hp at 6,500 rpm and 581.7 ft-lbs of torque at 4,900 rpm. Compression ratio was 11.0:1. Every part used in this build is listed. This brief information provides an example of how you can obtain a hefty power build by taking advantage of today's aftermarket components.

While any of a number of cylinder head configurations and induction systems may be used, this build features Trick Flow cathedral-port heads, a dual-carb tunnel ram intake manifold, and a pair of Holley Ultra 600 double-pumper carbs. To obtain more of an old-school appearance, adapters were installed to utilize Gen I small-block Chevy-style valve covers.

This build features a GM LS platform, bored and stroked to 408 ci, with a 4.030-inch bore and 4.000-inch stroke. A few of the notable elements include adapting Gen I small-block Chevy valve covers to LS heads and a conversion to carburetion, this time using a tunnel ram intake fitted with dual 4-barrel carbs.

The core block was salvaged from an iron LQ9 6.0L engine. The LQ9 is an LS platform that features a cast-iron block as opposed to an aluminum block. The entire core was purchased for a mere $400. The only items maintained from the core were the block and the rear engine cover. All remaining components are new. Most pieces were sourced from the performance aftermarket. The only GM parts purchased include the front-mounted camshaft position sensor, crankshaft position sensor, water temperature sensor, oil pressure sensor, rear-mounted plastic oil restrictor barbell, threaded block plugs, and plastic lifter trays.

The iron-block version was selected based on two factors: the relatively low cost of an engine core and the ability to easily accommodate an overbore. Once the core was torn down to a bare block, the block was hot-tanked, jet washed, and inspected for cracks. Luckily (although the original crank was roached) the block checked out fine, with no cracks and no evidence of ever having been remachined. The original cylinder bores were sonic checked for wall thickness, which revealed a minimum of .244 inch, safely allowing us to overbore final-hone to 4.0305 inches. Wall thickness must be checked before investing time in machining because of potential variation in wall thickness that commonly results from core shift on LS blocks.

Prior to block machining, I spent way too many hours smoothing out the block exterior to prepare the block for paint. Using die grinders,

Starting with a GM 6.0L iron block, this build featured a 4.0305-inch bore, 4.000-inch stroke, and 408 ci with 11.0:1 compression.

Unorthodox treatments included a Holley tunnel ram intake manifold with dual 600-cfm carbs and specially fitted small-block Chevy-style valve covers.

Basic Engine Dimensions

Original LQ9 Specifications (6.0L/364 ci)	
Bore	4.000 inches
Stroke	3.622 inches
Rod length	6.098 inches
Piston compression distance	1.331 inches
Block deck height	9.240 inches

Our Finished Specs (408 ci)	
Bore	4.0305 inches
Stroke	4.000 inches
Rod length	6.125 inches
Piston compression distance	1.115 inches
Block deck height	9.230 inches

deburring bits, Scotch-Brite pads, and abrasive wheels, the block was dressed by removing all factory casting flashings, high spots, pits, uneven surfaces, etc.

Block Machining

Gressman Powersports in Fremont, Ohio, machined our block. The block was bored and decked on a RMC CNC machine, which provides extreme precision and repeatability.

Prior to mounting the block in the CNC machine, the main cap register edges were deburred and installed with a set of ARP main studs. With the block mounted to the CNC machine, the digital probe obtained reference points, including main and cam bore locations, cylinder bore locations, and deck surfaces.

Our decks should have had a height of 9.240 inches. However, during a multilocation probe check on the CNC, these initially measured from 9.2324 inches to 9.2386 inches. The left bank was nominally higher than the right bank deck, so both were milled to a finished height of 9.230 inches, and thus both were made square and of equal distance from the crankshaft centerline. Based on our 4.000-inch stroke, 6.125-inch rods, and 1.115-inch piston compression height, the pistons protruded from the decks by .010 inch although use of .050-inch head gaskets compensated for this.

The cylinder bores, although spec'd at 4.000 inches, initially measured 3.9988 inches to 3.9997 inches and were slightly off-center compared to blueprint specs. Scott programmed the CNC cutter to open the bores up to a rough diameter of 4.025 inches and corrected bore centerlines in the process. Following the boring process, all cylinder top edges were then chamfered on CNC.

Once the block was accurized, the cylinders were honed on Scott's Rottler HP5 honing machine. Deck plates (along with crushed MLS head gaskets) were installed to the decks using the ARP head stud kit, snugging all nuts to 80 ft-lbs.

Cylinders were honed in three steps: honed to size (4.0305 inches) using 275 to 325 diamond stones with a motor load of 25 percent; four strokes with 500 stones at 20-percent motor load; and four strokes with plateau brushes at 20-percent motor load.

The main bore initially measured a fairly tight 2.750 inches, so Scott align-honed the mains to 2.7512 inches. This would theoretically provide about .002 inch of main bearing clearance, which would be closely checked during test fitting.

Four new GM threaded water jacket and oil plugs were installed on the outside of the block. These plugs feature straight metric threads. No NPT threads exist on the OEM block.

Cam Bearing Installation

The block's camshaft tunnel has different bore diameters. While locations 2 and 4 are the same, locations 1 and 5 are different, and location 3 is another diameter. The cam bearings in our Clevite kit are stamped with part numbers that must correspond to specific bore locations. Carefully inspect each bearing (noting its part number) and organize them on a clean workbench. To avoid mistakes, I label each with a black marker for easy selection (1, 2, 3, 4, 5). I used "1" to specify the front of the block.

Main Bearing Specifications			
Bearing Position	PN	Housing Bore	Bearing OD
1 and 5	SH-1995A	2.3472	2.3492
2 and 4	SH-2124	2.3276	2.3295
3	SH-2125	2.3079	2.3098

Our roller cam was a steel stick from Comp that was mated with our 1.7:1 rockers. Intake and exhaust valve lift is .624 inch. Duration at .050 is 243/251 degrees, with an LSA of 114 degrees.

Our Callies Compstar forged-steel crankshaft features a 4.000-inch stroke, 2.5585-inch mains, and 2.100-inch rod pins. It came already equipped with a 58-tooth reluctor wheel. This crank also features a keyed snout, allowing the use of either an OEM LS pulley or an aftermarket keyed balancer.

Camshaft

The Comp Cams' hydraulic roller, PN 54-462-11, was installed in this build. The lobe lift is .3670-inch intake and .3670-inch exhaust. Using 1.7:1 Harland Sharp roller rockers, our gross intake and exhaust valve lift is .624 inch. When test fitting and cam degreeing, the lift was determined to be right on spec. Duration is (as listed) 243 degrees intake and 251 degrees exhaust at .050 inch. The lobe separation angle is 114 degrees. Comp states that the operating range is 2,600 to 7,200 rpm.

Royal Purple Max Tuff assembly lube was used to coat cam lobes and journals. With the cam retainer plate in place, and with the Cloyes cam gear and Torrington bearing test fitted, we checked cam endplay at .0065 inch (generally, .005 to .010 inch is sufficient).

Crankshaft

A forged Compstar crank from Callies, PN APO31N, is the foundation of the rotating assembly. This crank has a 4.000-inch stroke, 2.5585-inch mains, and 2.100-inch rod pins. It was fitted with a 58-tooth reluctor wheel and rough-balanced. In addition, the crank has a rear key for the timing gear and oil pump drive and a front key to accommodate a keywayed damper. While LS OEM dampers are interference-fit only with no key, aftermarket dampers, such as our Fluidampr, are offered with keyways. So, with this crank, you have a choice of using either a keyed or no-key damper.

First, I test fitted the crank to in the align-honed block. The OEM spec calls for .0015 to .0078 inch in crank endplay and I measured endplay at .0025 inch. Main bearing clearance was measured at each journal at .002 inch. The crank main journal measurement is crucial and was 2.5585 inch while the installed main bearing inside diameter was 2.5605 inches. The crank endplay measurement fell within OEM specs, but the .0025-inch measurement was too tight, especially for a street performance vehicle. The number-3 main thrust bearings were removed and .001 inch of material was removed from the front and rear of the upper and lower thrust bearing faces. Measuring the bearing width at the thrust faces as a reference, I used an emery cloth to reduce the pair of upper/lower bearings, so I periodically remeasured for material loss, until .001 inch had been scrubbed from each thrust face. After reinstalling the crank and torquing all caps to spec, I measured the endplay at .0045 inch.

Main Cap Studs

Instead of using OEM-type main cap bolts, I opted for a set of ARP main studs, PN 234-5608.

The studs are 10 mm, with 2.0 threads at the top (engaging into the block) and 1.25 threads at the lower end (for the nuts). The LS main studs feature two different lengths (10 studs measure 4.450 inches in length and 10 measure 4.773 inches in length).

The shorter studs are installed in the outboard locations and the longer studs are installed in the inboard locations, closest to the crank centerline. The thread depths in the block result in the shorter exposed outboard studs projecting farther out to accommodate the windage tray and windage tray nuts.

Finger-tighten the studs to the block. *You must not excessively tighten the studs to the block.* With caps and lubed bearings in place, use ARP assembly lube to coat the exposed stud and nut threads. Final torque of the inboard nuts is 50 ft-lbs, and final torque of outboard nuts is 60 ft-lbs. The 8-mm side bolts (each cap features a right-hand [RH] and left-hand [LH] side bolt) are torqued to 20 ft-lbs. *You should always tighten the primary 10-mm stud nuts first, followed by tightening the side bolts.*

Crank Balance

Jody Holtrey at Medina Mountain Motors in Creston, Ohio, balanced the crankshaft. First, each piston, each pin, each rod, each pair of rod bearings, each pack of rings, and each pair of pin locks were weighed. Not surprisingly, no corrections were needed on the pistons or rods. Both JE's and Callies's components serve as ideal examples of the high level of precision manufacturing in today's performance aftermarket. Each piston weighed in at 390 grams. Each pin weighed in at 106 grams. Each rod's big end was at 452 grams and each rod's small end was at 174 grams. With the quality that I've come to expect from these manufacturers, there's really no need to weigh each piece of a set, but it never hurts to take the time to check.

The crank initially spun at less than 40 g heavy at each end. After weight relieving by drilling counterweight edges, Jody achieved –0.7 g to –0.4 g, which is within the ideal range for street applications . . . basically anything within 4 or 5 grams is fine.

Bobweight

Listed here is our bobweight info (all in grams):	
Piston	390
Pin	106
Rings	41
Locks	3
Small end	174
Big end	452
Bearing (44 x 2)	88
Total	1,706
1/2 Total	853 (50 percent)
Plus oil	4
1/2 bobweight	870 g (51 percent)

Connecting Rods

Callies's Compstar H-beam forged-steel rods, PN CSC6125 DS2A2AH, were installed and feature a center-to-center length of 6.125 inches. The pin end width is reduced to accommodate forged side relief (FSR) pistons. Some FSR pistons have pin towers relocated inward, and therefore, the pin end of the rod needs to be narrowed. The Compstar rods accommodate stroker builds and feature optimized profiling and bolt placement to increase camshaft and oil pan rail clearance. During test fitting, no clearancing was required.

These rods also feature strengthening gussets added to the cap surrounding the relocated bolt spot faces fitted with ARP 2000 rod bolts. A chamfered parting edge prevents material from scraping the during installation. The polished beam edges reduce the risk of stress risers. A Sunnen Krossgrinder was used for finish honing to achieve tolerance of less than .0002-inch maximum out-of-round, taper, hourglass, or barrel shape. The bearing housing surface finish has an Ra of 9 to 12.

The 7/16-inch ARP 2000 rod bolts feature a shank length of 1.450 inches and Callies specifies 75 ft-lbs of torque or a measured stretch of .004 to .005 inch.

After applying ARP assembly lube to the threads and underside of the bolt heads during final installation, I indexed my stretch gauge to each bolt's free length, then torqued to 70 ft-lbs and checked with my stretch gauge, continuing to tighten until each bolt achieved a .004- to .005-inch stretch.

I used Clevite CB-663HNK on the 2.100-inch rod pins and measured initial rod bearing clearance at .0019 inch, which was way too tight. I kept the standard-size CB-663HNK upper bearings, but I installed 1X lower bearings (CB-663HXK) for a bearing clearance of .0024 inch.

Connecting rod sideplay was measured during test fitting, with each pair of rod big ends featuring .014 inch of sideplay. It was obvious that Callies held very precise tolerances on both the Compstar crank and rod set.

Piston Rings

JE end gaps for rings are as follows. This is based on bore diameter and intended application (all listed in inches):

Piston Ring Specifications			
Application	Top	Second	Oil Rail
High-performance street	Bore x .0045	Bore x .0050	Min .015
Street moderate turbo/nitrous	Bore x .0050	Bore x .0055	Min .015
Late-model stock	Bore x .0050	Bore x .0053	Min .015
Circle track/drags	Bore x .0055	Bore x .0057	Min .015
Nitrous race only	Bore x .0070	Bore x .0065	Min .015
Blown race only	Bore x .0060	Bore x .0060	Min .015

Whenever you deal with a very short piston compression height, the wrist pin bore may intersect with the floor of the oil ring groove. For this piston design, the piston maker supplies oil ring support rails. The support rail installs prior to the oil ring package, providing a fill support for the oil rings at each side of the wrist pin bore. The support rail features a small male bump, which must be positioned at one of the ring groove floor openings. This bump prevents the support rail from accidentally rotating, ensuring that the support rail's end gap does not become exposed.

For this particular high-performance street engine with a 4.030-inch bore, a top ring gap at .018 inch was required, and the second ring required a .020-inch end gap. Each ring is specifically fitted to each individual cylinder bore. Cylinder diameter may vary slightly from bore to bore, and as a result, it affects end gap. Once a ring has been fitted to a specific cylinder, you need to make sure it's used for that respective cylinder.

After an initial test fitting was conducted, I measured an end gap of .002 inch for the top ring while the second rings had zero gap. Using a Summit bench-mounted rotary diamond-wheel ring filer, I filed the top rings to arrive at a gap of .018 inch and I filed the second rings to .020 inch. One I had achieved the desired end gap, I used a fine ignition file to carefully deburr the filed edges. The oil ring rails were checked at .023 inch for a minimum allowable gap is .015 inch.

Since the 1.115-inch compression height of the pistons brings the pin bore into the oil ring groove, you must first install a support rail at the bottom of the oil ring groove. This spans the gap where the pin bore intersects the ring groove, restoring oil ring support in these two opposing spots. When installing the support rail, the rail's male dimple must face downward, centered over one of the pin bores. This prevents the support rail from rotating too far and eliminates any chance of the support rail's end gap entering the open span over the pin bore.

Pistons

JE's forged asymmetric pistons were installed. These feature a wider skirt outboard (thrust side) on the right bank and inboard on the left bank. In addition, the opposite side has a narrower skirt, and the pin centerline has been relocated about .020 inch toward the thrust side to compensate for center balance. These flattop pistons (PN 311979) feature valve pockets for 12- to 15-degree valve angles. Piston compression distance measures 1.115 inches, while skirt diameter is 4.025 inches. JE recommends .005-inch piston-to-wall clearance, and this particular build has a 4.030-inch bore diameter. The full-floating piston pins are .927-inch diameter and are 2.250 inches long, and .073-inch single wire locks secure them. Each piston tips the scale at just 390 grams.

JE forged pistons feature 1.115-inch compression height, asymmetric skirts, and skirt antifriction coating.

Piston Widths

Ring Groove Widths	
Top	.048 inch
Second	.060 inch
Oil	.149 inch

Land Widths	
Top	.235 inch
Second	.155 inch
Oil	.085 inch

Radial Depths	
Top	.162 inch
Second	.177 inch
Oil	.172 inch

Timing Set

When it comes time to buy a timing set, you need to be aware of the differences in the camshaft gear for the specific version of the LS engine. General Motors has four different LS camshaft gears.

- Single-bolt 4X has a camshaft with a single center bolt (gear to cam), and it has been installed in some 2007–2009 LS2 and LS3 engines.
- Three-bolt 4X was installed on camshafts with three-bolt mating (gear to cam). These were standard equipment on the 2006 LS2 Corvette, 2006–2009 LS7, and 2009–2010 LS9.
- Three-bolt 1X was installed in 2005 LS2 Corvette, 2005–2006 LS2 GTO, SSR, and Trailblazer.
- Three-bolt timing set without camshaft position sensor reluctor teeth on the cam gear was installed in 1997–2004 LS1, LS6, and other Gen III engines.

The camshaft position sensor location and the number of teeth on the crankshaft reluctor wheel need to be considered. A 1X cam gear is required if the crank reluctor has 24 teeth and the cam position sensor is located at the top rear of the block. A 4X cam gear is needed if the crank reluctor wheel has 58 teeth and the cam sensor is located on the engine front cover.

Apply a small amount of lube to the crank snout prior to installation of the crank gear. Place the gear onto the crank's snout and align the desired timing keyway to the crank's key. I then used a thick-wall aluminum driver with hollow entry to tap the gear into seated position. Never hit the gear directly with a hammer or other solid object.

The Cloyes Hex-Adjust cam gear is adjustable within a range of 4 degrees advance or retard, thanks to the eccentric bushing that is turned with a hex wrench.

The cam gear should be mounted to the cam at this point, but do not tighten the cam gear bolts. Use a 1/4-inch hex wrench to adjust cam timing, which is at the eccentric bushing. If the cam bolts are already tight when you try to change cam timing with the hex adjuster, the hex adjuster will split.

Once cam gear timing has been adjusted, tighten the cam gear bolts (with threadlocker on the threads) to 26 ft-lbs (for three-bolt design). I used ARP's cam gear bolts, PN 134-1003 (8 mm x 1.25 x 25 mm) with a dab of 242 threadlocker.

If using a single-bolt gear-to-cam style, the center bolt is a TTY (torque-to-yield) bolt. If using an OEM TTY bolt, always use a new TTY bolt. Apply thread locker and initially tighten to 55 ft-lbs, followed by an additional 50 degrees.

Timing Chain Damper

GM did not equip the LQ9 block with a timing chain damper provision. An LS2 block, for example, has a stationary composite damper, and two 8-mm bolts are used to mount it to the block between the cam and crank gears. Trick Flow's LS timing chain damper adapter bracket (PN TFS-306756000) allows you to quickly mount an LS2-style chain damper to an LQ9 block. Install the cam retainer plate to the front of the block and tighten to finger strength. In this case, I used one of the retainer plate bolts from my ARP set (P/N 134-1002) that has 8-mm x 1.25 x 20-mm hex-head bolts. I applied a drop of threadlocker before installation.

Use the supplied 8-mm x 1.25 x 20-mm button-head bolts with threadlocker to install the Trick Flow adapter bracket at the remaining three retainer plate bolt locations. A 10-mm socket wrench for the ARP bolt and a 5-mm hex bit are required to tighten the button-head bolts. All four bolts need to be torqued to 15 to 18 ft-lbs.

With a replacement plastic damper, use a socket to drive out the original powdered metal bushings from the damper's two bolt holes. Install the plastic damper to the bracket using the supplied two 8-mm x 1.25 x 10-mm flanged button-head bolts and place threadlocker on the threads. Torque the bolts to 15 to 18 ft-lbs. In some cases, damper bushing holes are different diameters, so in case your damper holes are too large for the button-head bolt heads, the kit includes a pair of aluminum spacers. You must use the supplied button-head bolts to provide needed clearance, so don't try swapping to taller-head bolts, and drilling of the block is not required. This procedure takes about five minutes of time.

Crank and Cam Position Sensors

All the OEM sensors featured in the fuel-injected version are not required to operate the ignition system on any of the LS platforms. However, the crankshaft and camshaft position sensors are required. These sensors deliver information about the crank position and cam position for the MSD ignition control module to manage coil-to-spark plug firing.

The crankshaft position sensor mounts to the right rear side of the block, inline with the crank's toothed reluctor wheel. The sensor is sealed with an O-ring. A mounting tab secures the sensor to the block using a single 6-mm x 1.0 x 15- to 20-mm bolt. A 58-tooth reluctor wheel has been installed on our Callies crank, and therefore the compatible crank sensor is GM PN 12585546. An 8-mm x 1.25 x 15-mm retaining bolt is torqued to 18 ft-lbs and holds the sensor in place.

According to the version of the LS block, the camshaft position sensor mounts either at the top rear of the block or onto the front timing cover. A rear-mounted cam sensor was originally installed on our LQ9, and it's compatible with a 24-tooth crank reluctor. Designs following the LS1, LS6, and early LS2 feature 58-tooth crank reluctors that have a cam sensor installed on the timing cover. Cam sensors mounted at the rear of the engine send cam position information through a reluctor that's built into the rear of the cam and near the cam's rear journal. Front-mounted cam sensors require the use of a 4X cam gear and pick up their cam position signal from the camshaft sprocket gear. I opted for the 58-tooth crank wheel and the late-LS2-style front-mount cam sensor because this particular version delivers increased ignition timing accuracy.

I installed the cam sensor up front, so I needed to plug the OEM cam sensor hole in the rear. The upper hole sits flush with the top cover's gasket surface, so there's no need to plug it because it's simply open to air underneath. When the OEM cam sensor is placed in the bottom hole, it seals in the block. Thus, when the sensor is moved, it needs to be plugged to avoid an oil leak. Fortunately, a 1/2-inch NPT thread fits in the lower hole. I tapped this hole at the beginning of the build, so it was done before the block was cleaned and painted. During final assembly I mounted a 1/2-inch NPT plug with a female hex and thread sealer was also applied. The top hole was left open, and it provides easy access for a hex wrench.

Water Temp Sensor

All cylinder heads have a 12-mm x 1.5 threaded hole for mounting the OEM water temperature sensor. For this project, I mounted a GM sensor (PN 12608814) in the sensor hole on the left-side head. Also, I ensured that the threads were coated with Teflon thread sealer. To plug the same hole on the right-side head, I simply installed a short 12-mm x 1.5 hex-head bolt (with crush washer). Length isn't a major consideration, so I installed a 20-mm-long bolt. However, you need to plug the unused water hole. You cannot separately buy the GM plug that's installed on OEM right-side heads. GM only offers the plug as part of a head, so you'd need to buy a cylinder head just to get this plug.

You should always test fit the exhaust headers to the heads to ensure there's adequate clearance at the water temp sensor (and/or plug) and the header flange. For example, the water temp hole is positioned too close to the adjacent exhaust port on the World Warhawk heads, and with

Rather than using a stock front cover, a Comp Cams cover provided extra depth for chain clearance and an adjustable-position bore for the front-mounted LS2 style cam position sensor.

some headers, a bit of material from the leading edge of the header flange needs to be ground off for clearance to the sensor. The Trick Flow heads do not present a header clearance issue.

Both left- and right-side cylinder heads are identical, so the threaded hole for the temp sensor needs to be plugged on the right-hand head. OEM right-hand heads have a threaded plug, but GM does not offer it individually. The fix is easy. Simply use a 12-mm x 1.5 x 15- to 20-mm hex-head bolt. Install a crush washer and apply thread sealant and install. You don't need the unique GM plug.

Windage Tray and Oil Pan

The OEM windage tray extends over the exposed outboard main stud tips. The tray is stamped for rear orientation, so the tray cutout is designed to align on the right side for the dipstick. This build features an increase at 4.00-inch stroke, and as such, the connecting rods hit the OEM windage tray. If you're going to install the OEM tray, the tray needs to be spaced away from the main caps by about .230 inch. Use small-diameter hardened steel washers are ideal for this and stacking two 10-mm washers with a thickness of about .115 inch onto each outboard main cap stud tip will suffice. The zinc-plated serrated-surface locking nuts supplied with the ARP main stud kit should be used and all windage tray nuts should be torque to 18 ft-lbs.

The Holley aluminum oil pan has a sump cover baffle plate, and it extends over most of the sump area. Therefore, you just install the Holley plate with no need to modify the OEM tray.

If you install the Holley pan but want to use the OEM windage tray, you need to follow the previously mentioned spacing instructions. However, you also need to verify that the washers are small enough in outside diameter to accommodate the rear of the Holley oil pan. Hence, if the washers are too large in outside diameter (OD), the pan's bolt holes won't align with the block's pan rail holes. Washers with a 19-mm OD are ideal for this. In addition, several aftermarket windage trays provide enough clearance for stroker engines. However, using spacer washers, you can easily obtain enough rod clearance.

By closely examining the Holley pickup tube support bracket, you can see it seats at the number-4 right-hand windage tray stud location. It needs to sit flat. If it does not, the curved corners inboard of the bolt hole may be preventing the bracket from seating flush. If this is indeed the case, a die grinder can be used to level out the two radiused corners until the bracket is permitted to seat flat. The Holley oil pump pickup attaches to the oil pump with an included O-ring. Spread clean engine oil over the O-ring, and torque the single 6-mm mounting bolt (tube to pump) to 106 in-lbs. Mount the pickup support bracket to the left-hand number-4 main stud and torque to 18 ft-lbs.

The supplied 1/4 x 20 x .50–inch socket head cap screws fasten the Holley baffle tray to the pan. Apply threadlocker, and torque to 8 to 10 ft-lbs. A set of ARP stainless LS oil pan bolts secure the cast-aluminum Holley pan, and a pan gasket from our Mahle-Victor gasket set provides the proper seal. Place a 20-mm-long strip of RTV at the front of the pan rails where the front cover comes together with the block.

Due to the increased crank stroke, it was necessary to shim the windage tray in order to gain rod clearance.

Torque the pan-to-block as well as the pan-to-front-cover bolts to 18 ft-lbs. The pan-to-rear-cover 6-mm x 1.0 x 30-mm-long bolts need to be tightened to 106 in-lbs. When test fitting the oil pan, you need to ensure that the pan is properly aligned. The rear of the pan needs to be flush with the rear of the block to accommodate the transmission bellhousing. The Holley LS oil pan sump has a 5.5-quart capacity and a total capacity of 6 quarts with a stock-size filter.

Oil Cooler Adapter

An engine oil cooler provides an extra measure of engine protection and is absolutely necessary for some applications. You need to verify that it will fit your particular vehicle before installing an engine oil cooler. I mounted a well-designed Lingenfelter billet-aluminum adapter block (PN L300025297) to the Holley oil pan. The adapter boss is found above the oil filter base. This adapter is secured to the pan with a pair of stainless-steel 6-mm x 1.0 x 35-mm SHCS (socket head cap screws). The inlet/outlet fittings are male –10 AN 37-degree fittings, ready to accept –10 hose ends. The adapter kit includes the adapter block, a pair of mounting screws, and an OEM-type double-O-ring seal.

The –10 AN male fittings provide a path in and out of the oil cooler while the adapter features a bottom-position 1/8-inch NPT female port for turbocharger or supercharger oil feed applications. If you don't need to use this, an 1/8-inch NPT plug is provided to plug it. A side-position oil temperature sensor port with 12-mm x 1.5 female threads is another feature. If you do not need to use this port for a sensor, the kit includes a stainless-steel 12-mm x 1.5 hex-head plug and copper crush washer to close it off.

Apply a medium threadlocking compound to the 6-mm mounting bolt threads, and torque these bolts to 106 in-lbs (12 Nm) when installing the adapter. Apply a medium threadlocking compound to these threads whether installing an oil temperature sensor or the 12-mm crush-washer-equipped thread plug. You can use the oil temperature sensor port when adding an oil temperature sensor to a system that didn't feature an oil temperature provision. The temperature is measured at this port before it reaches the cooler.

Don't use this type of adapter when using an external engine oil filter. This adapter is compatible with a cooler and does not feature internal bypass. The –10 fittings must be attached to and from an oil cooler or looped together. If you simply cap off the –10 fittings, it won't produce any oil pressure.

If an engine oil cooler is planned, a Lingenfelter oil adapter mounts directly to the oil pan's port locations with no modifications. This replaces the OEM bypass cover that is used if no oil cooler was installed.

As for fitment to LS-equipped production vehicles, this adapter is designed to fit 1997–2004 C5 Corvette, 2005–2009 C6 Corvette, 2004–2006 Cadillac CTS-V, 1998–2002 LS1 Camaro and Firebird, and 1999–2007 4.8, 5.3, 6.0, and 6.2L GM CK trucks. This adapter does not fit 2004–2006 Pontiac GTO. Of course, if you're adapting an LS engine to a street rod, muscle car, or other swap, check for clearance (as with any swap or mod).

Oil Pump

For this project, a Melling Select performance pump (PN 10296) with a high-pressure relief spring was installed. Install the supplied blue spring for a lower pressure system and tighten the valve plug to 106 in-lbs. Before mounting the pump, simply slip the supplied oil pump drive onto the crank snout drive; it comes with the timing set. If the installation is difficult, stop and check for burrs on the key or keyway. Then spread a bit of oil to the inside of the drive and slide it fully onto the snout until it bottoms out against the crank gear. One side of the pump drive gear remains slightly chamfered so it more easily slides down onto the snout. It will leave 1.250 inches of exposed crank snout for the crank damper once it has completely seated.

After applying some lube to the outer tooth area of the crank-snout-mounted oil pump drive gear, carefully slide the oil pump into place, so the drive and driven gears are engaged.

Before final-tightening the pump mounting bolts, nudge the pump body around to center the pump's main bore with the crank centerline. If there is too much play, you need some special aftermarket alignment shims. However, you can precisely measure radial distance using a caliper to measure the play. A tiny bit of play should be present in the pump body mounting holes to properly center the pump. You can't just simply bolt the pump onto the block, so closely monitor the pump orientation so you obtain the ideal centering position. Secure the pump with four 8-mm x 1.25 x 30-mm bolts, and tighten to 18 ft-lbs.

Once the pump has been installed, apply lubricant to a new oil pump pickup tube O-ring and install the O-ring to the pickup tube. Insert the pickup tube into the pump exit, being careful not to damage the O-ring.

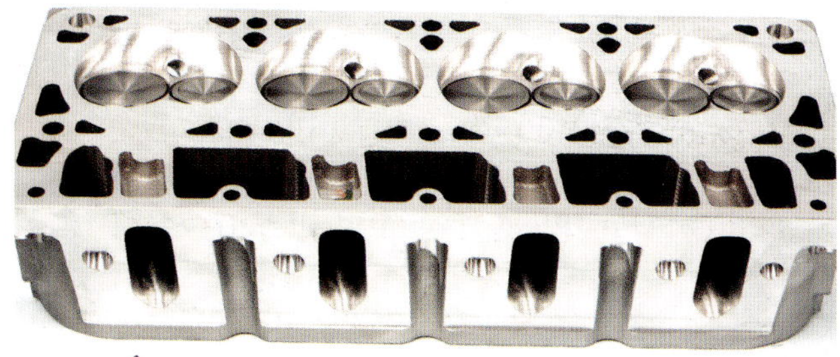

The cylinder heads selected for this build are Trick Flow's TFS-3061T001-C02, featuring LS2 style cathedral intake ports, 225-cc intake runner volume, and 65-cc combustion chamber volume. These heads are fully CNC finished and fully assembled.

Both main caps and cylinder head decks were fitted with ARP studs.

Cylinder Heads

A pair of Trick Flow GenX heads (PN TFS-3061T001-C02) was bolted to the block. These feature LS2-style cathedral ports and a combustion chamber volume of 65cc. Intake runner volume is 225 cc while exhaust runner volume is 80 cc. These fully assembled heads feature CNC-machining in ports and chambers. The intake valve diameter is 2.055 inches and exhaust valve diameter is 1.575 inches while valve spring diameter measures 1.300 inches, fitted with titanium retainers.

Cylinder Head Studs

I also used an ARP stud set (PN 434-4316) with 12-point nuts and hardened washers. A complete array of all studs was needed for both heads. Our LQ9 block requires an unequal-length set, and each head needs eight 11-mm studs at an overall length of 185 mm, two 11-mm studs at an overall length of 130 mm, and five 8-mm studs at an overall length of 65 mm.

If you're dealing with a 2004 or later LS block, the ARP stud kit (PN 434-4317) includes 11-mm primary studs and these are all 130 mm in overall length.

The ARP primary studs measure 11 mm x 2.0 for lower thread for block engagement and have an upper thread size of 11 mm x 1.25.

We cleaned and chased all block threads as we prepped the block. It's an important step for an iron block because rust can quickly set in on female threads. During block machining and test fitting, we've handled the studs, nuts, and washers, so each piece was thoroughly washed and cleaned. The studs were threaded into the block at finger-tightness. I lightly lubed the lower threads with oil, threaded them in until they bottomed out, and then backed off just a tad to avoid any harsh preloading. The Mahle-Victor MLS head gaskets and head were put in place. ARP assembly lube was used to coat both sides of each washer, female threads, undersides of the nuts, and the upper stud threads. With washers and nuts in place, the primary nuts were torqued to a final spec of 80 ft-lbs. The four inboard 8-mm stud nuts were tightened to 25 ft-lbs.

Cylinder Head Tightening Sequence

For the 10 primary 11-mm fasteners:

1	Upper row, center
2	Lower row, center
3	Upper row, left of center
4	Lower row, right of center
5	Upper row, right of center
6	Lower row, left of center
7	Lower row, far left
8	Lower row, far right
9	Upper row, far right
10	Upper row, far left

For the five 8-mm fasteners:

11	Center
12	Right of center
13	Left of center
14	Far right
15	Far left

Lifters

I opted for Comp Cams' "short race" hydraulic roller lifters (.842-inch diameter) for this build. Plastic lifter "buckets" guide an LS engine's lifters. The four bucket units capture and guide a set of four lifters. Roller lifters must be prevented from rotating in their bores to keep the lifter's roller bearing in plane with the cam lobe. Instead of using dogbones or bridged lifters, GM installed composite lifter trays/buckets that have female flats to engage and guide each lifter. A set of four lifters is inserted into a bucket, and then the assembly is dropped into the block with the lifters entering their respective lifter bores. A single 6-mm x 1.0 shouldered hex bolt secures each bucket assembly to the block. You must use the OEM-style shouldered bolts to properly center each lifter bucket bolt hole to its threaded hole in the block. The bolts are torqued to 125 in-lbs. When the lifter assemblies have been installed in the block the lifters are still held up, away from the cam lobes, so at this stage, you simply insert a pushrod and pop each lifter down onto its cam lobe.

This lifter bucket design makes future camshaft removal/swaps much easier. The removal process is quick and convenient. Remove all rocker arms, and then rotate the crank two full revolutions. With no pressure on the lifters, the lifters are pushed up and "held" in their buckets, which moves the lifters out of the way of the cam. First remove the crank damper, water pump, front cover, and cam gear. Then you can slide the cam out of the cam tunnel and slip a new cam in. Once the camshaft has been installed, a pushrod is used to pop the lifters out of the held position in the lifter buckets. Lifter, intake manifold, or head removal is not required.

Our lifters are Comp's short race hydraulic rollers, featuring a 0.842-inch diameter.

The lifters have a click-in snug fit in the buckets. The lifters are pushed fully upward into the buckets and are fairly reliable. However, the possibility of a lifter falling out of place and dropping into the cam bore area needs to be eliminated. Hence, when the lifters have been raised up into their buckets, you need to place a 24-inch-or-longer 5/16-inch metal rod into each oil galley, which is above and at opposite sides of the cam bore. These temporary rods serve as a fail-safe. If one or more lifters become dislodged from their buckets, the rods prevent the lifters from traveling far enough to cause a problem. Once the cam is reinstalled, remove the rods.

Rocker Fitting to the Heads

The refined design of the Harland Sharp PN SHLS17 aluminum full-roller rocker arms offers superior strength and rigidity. These rocker arms also offer slightly chamfered edges to further reduce arm weight. Our rockers feature a 1.7:1 ratio and are mounted with supplied support stands and 8-mm SHCS (socket head cap screws), tightened to 22 ft-lbs. Always check rocker arm clearance to the cylinder heads, as head designs can differ in terms of clearance issues. On the Trick Flow heads, clearance simply isn't an issue . . . bolt them on and proceed to measure for pushrod length.

Harland Sharp 1.7:1 forged full-roller rockers and custom-length Trend pushrods completed the valvetrain.

Pushrods

Using an adjustable-length pushrod checking tool, we determined an ideal pushrod length at 7.425 inch. Our Comp Cams Hi-Tech pushrods feature 5/16-inch diameter, 7.425-inch length, and a wall thickness of .080 inch.

Prior to installation, I ran a pushrod cleaning brush with a fast-drying brake-cleaning solvent through each pushrod several times, followed by a hot soapy water wash and rinse, followed by a compressed air blow-through. Each pushrod received a coating of Royal Purple Max Tuff assembly lube at each tip.

Front Cover

Whether you call it a front cover or timing cover, you need to verify that the gasket surface on the block is clean and dry. Assemble the metal-core cover gasket (from our Mahle-Victor gasket set) to the front cover by inserting the cover bolts through both the cover and gasket (this will hold the gasket in place). Place the cover in position on the block face and begin to finger-thread all the 8-mm bolts. Once all bolts have been started, push the cover flush to the block. Begin to tighten all bolts in a crisscross manner (to spread the load equally across the cover) until all bottom out finger-tight only.

Rotate the block upside down. Place a straightedge along both the oil pan rails and the bottom of the front cover. Adjust the cover position so that the bottom surface is flush with the pan rail. Then tighten a couple of bolts to maintain that flush position.

Torque all bolts (again, using a crisscross pattern) to a final value of 18 ft-lbs. I had a choice of front covers for this build that would accept the front-mounted camshaft position sensor: a new OEM LS2-style cover or a new Comp Cams cover, PN 5496, and I chose to install the Comp cover.

Water Pump

The water pump is secured to an LS block face with two metal-core gaskets and four bolts. I opted for a polished finish and top-of-the-line Meziere electric pump (PN WP319U). For the thermostat neck housing, I could have installed an OEM unit, but I decided to install a Meziere polished billet-aluminum water neck, PN WN0019U.

I installed the pump to the block using four 8-mm x 1.25 x 115-mm stainless-steel socket head cap screws, tightened to 11 ft-lbs on the first pass, followed by 22 ft-lbs.

Crank Damper

I selected a damper with integrated pulley grooves from Fluidampr, PN 740111, to provide ideal balancing for the application. This model features an OEM pulley diameter of 7.5 inches, and it has the same size and belt groove design as the LS1 Camaro/Firebird. Also, the damper has a diameter of 6.250 inches. Keep in mind, the OEM Chevy/GM dampers are a press-fit only, so there is no registering key. While the key obviously prevents damper rotation relative to the crank, it permits installation of a timing pointer for number-1 cylinder reference.

The damper had an interference fit of .0015 inch. Our crank snout measured 1.4820 inches in diameter, and our damper bore measured 1.4805 inches.

Our Fluidampr assembly featured a modular balancer and pulley system.

A specialized installation tool is required for properly installing an LS damper. This tool has a male thread of 16 mm x 2.0 to engage the crank snout threads. Typically a damper installation kit includes a 16-mm x 2.0 nosepiece. While specialty threaded inserts are available for this depth, I cheated and built an installation tool. I used a 10-inch-long Grade 8 16-mm x 2.0 piece of all-thread and I machined a steel mandrel on my lathe that features a 16.5-mm pass-through hole. In addition, I also grabbed a 2-inch-diameter flat washer and a caged bearing from my Moroso damper installer kit. If you need to remove the pulley from the damper, upon reassembly, the three 12-point mounting bolts should be torqued to 37 ft-lbs, with thread locker applied.

Timing Pointer

The Comp Cams front cover includes a provision for attaching a Chevy big-block timing pointer. I installed a TCI adjustable timing pointer (PN 871005). A pair of included 1/4-inch x 20 x 1/2-inch socket head cap screws fastens the timing pointer adapter to the front cover. When I installed them, I used a drop of threadlocker on each screw

The Comp front cover features a provision to easily mount an adjustable timing pointer.

and torqued both to 106 in-lbs.

The timing pointer mounts to the Comp adapter with a pair of included stainless 1/4-inch x 20 x 1½-inch SHCS, and these were torqued to 106 in-lbs. The front cover is compatible with the stock OEM cam sensor, but it won't accommodate the OEM cam sensor bracket. It does not have a provision on the Comp cover to secure the bottom of the bracket. However, the OEM front cover has a 6-mm x 1.0 threaded hole for the bracket bottom. Granted, a blank boss is incorporated into the Comp cover and it aligns with the cam sensor bracket's bottom locating dimple. It's located just above the bracket's lower bolt hole. It is possible to drill out the dimple to accept a 6-mm bolt, and to drill and tap the cover boss for 6 mm x 1.0 or 1/4 x 20.

However, use of the OEM cam sensor bracket interferes with the timing pointer, requiring spacing the pointer out from the adapter by about 1/4 inch. I installed the sensor to the cover with a 6-mm x 1.0 x 15-mm SHCS and flat washer. To secure and protect the dangling sensor wire harness and connector, I shielded the sensor wires with a length of .188-inch-inside-diameter braided loom.

To provide a timing reference, I installed a trimmed-down adhesive timing tape (for the 7.25-inch-diameter Fluidampr pulley) to the pulley's .250-inch-wide center rib. With the number-1 piston at TDC in the firing position, I aligned the tape's 0 mark to the pointer needle. The needle is adjustable to fine-tune it for perfect alignment to the tape.

Valley Cover

The LQ9's original cast-aluminum valley cover is unsightly and it has two big bores for valley-mounted knock sensors. I didn't plan to use knock sensors so I didn't want those two ugly holes visible on the cover. Therefore, I started out the fabrication process using a piece of flat stock that's 6 inches wide x 20 inches long and I wound up with far better-looking valley cover using 1/4-inch-thick T6061 aluminum.

The OEM cover measures exactly 6 inches wide. I cut my raw piece of flat stock to length. I then contour cut the rear so it cleared the OEM oil pressure sensor location. The LQ9's cam sensor is located at the block's top rear, so I used an LS2 front cover that features a front-mounted cam sensor. In addition, I also had to plug the original cam sensor bore in the block. I ended up making the new valley plate marginally longer at 19⁷⁄₁₆ inches. By making it a bit longer, it saved me from making a separate plug/cover for the cam sensor hole. The top surface of the LQ9 cam sensor location is flush with the mounting surface of the valley cover.

Installing the new valley plate then required 11 bolts instead of the original 10. These 10 bolts were needed for the standard locations plus one bolt at the cam sensor bracket hole. All bolts for the valley cover are ARP 12-point polished stainless steel, 8 mm x 1.25 x 20 mm (snugged to 18 ft-lbs). My custom valley cover didn't have the original milled gasket ring recess, so I used a thin bead of RTV to seal the cover. My favorite choice for this type of application is Permatex's The Right Stuff. This high-density black RTV provides an excellent seal. To keep the valley cover looking professional, I first masked the valley plate edges and surrounding adjacent block surfaces, and after that, I torqued the bolts to 18 ft-lbs. I then carefully removed any excess RTV. When the masking tape was removed, the result was a clean installation with no RTV exposed.

Intake Manifold

The intake manifold, Holley's LS Modular Hi-Ram carbureted unit, PN 300-226, is designed for use with LS1/LS2/LS6 cathedral-port heads. The intake manifold assembly I chose consists of two parts: the lower base PN 300-226, along with the upper carb base PN 300-216. This top allows the mounting of two 4150 series side-mount carbs. Holley offers the Hi-Ram in several configurations, with lower bases available for cathedral-port heads or for LS3/L92 rectangular-port heads and tops that accept 4150 series carbs, 4500 series carbs, or EFI throttle bodies.

During test fitting, I port-matched the intake to the heads, which required very minimal mods. The alignment and dimensions were already very close straight out of the box. The lower manifold decks were also machined ultrastraight, with nicely executed mating from front to rear on both heads. The lower manifold base is sealed via supplied O-rings that insert into machined grooves in the manifold decks. According to Holley, the Hi-Ram manifold is designed for use on naturally aspirated and forced-induction engines in the 6.0 to 7.0L range, accommodating power up to the 7,000- to 8,000-rpm range.

Port size is 2.49 inches high x 1.21 inches wide. Runner length is 6.58 inches. The as-cast runner cross sectional area is tapered from 4.25 inches square to 2.53 inches square.

Intake Manifold Fit

The angle of intake manifold needs to be mated to the head intake decks. Once that has been accomplished, you need to verify the intake and head ports are precisely aligned, and if not, you need to gently blend the intake manifold ports to their respective ports on the heads. Our Trick Flow heads' manifold-to-head bolt holes are aligned perfectly, so no adjustment

The Holley Hi-Ram intake improves top-end power and allows mounting either a single carb or a pair of carbs.

was necessary at those locations. Ten 6-mm x 1.0 fasteners secure the intake manifold to the heads, and a silicone bead provides an excellent seal. A light coat of oil or lithium grease was applied to the O-ring seals and then the O-rings were inserted into the seal groove that surrounds each cathedral intake port. The lube helps hold the seals in place during handling.

Once the manifold was correctly positioned, I finger-tightened the SHCS in the remaining locations. I then removed the four studs and finished placing the SHCS. I applied a small dab of ARP lube to the screw threads before installing. All 10 fastener locations are initially tightened to 50 in-lbs and the torqued to 106 in-lbs. As logic dictates, you need to follow Holley's recommended tightening pattern, and it specifies starting at the center and working your way outward. The upper plenum fastens to the intake manifold with 12¼ x 20 x 7/8–inch SHCS. Before mounting the upper plenum, place the round-profile bead O-ring seal in the machined groove of the intake manifold's plenum flange. The intake manifold flange has nonthreaded holes, while the upper plenum holes are threaded. As a result, the screws are inserted from the bottom of the intake's plenum flange. Lightly lubricate the threads and install, tightening to Holley's spec of 130 in-lbs. I used a set of stainless SHCS and washers from Totally Stainless.

Carburetors

Since I was using a tall, stick-out-of-the-hood intake setup, I opted for Holley's top-of-the-line Aluminum Ultra HP carbs, PN 0-80801RD, sized at 600 cfm. These carbs deliver 20 percent more fuel bowl capacity and full adjustability. This includes emulsion bleeds, power valve channel restrictors, and idle feed restrictors, with power valve channel restrictors machined lower in the metering block for improved fuel delivery to the power valve circuit. Basically, it's a leading performance design coupled with ease of adjustment/service features resulting from years of development. While I chose a pair of 600-cfm carbs for this LS build, the Ultra HP is also offered in a 650-, 750-, 850-, and a true 950-cfm version.

A pair of Holley all-aluminum Ultra HP 600-cfm double-pumper carbs feed our radical induction system.

Fuel Plumbing

I decided to plumb the carbs with 3/8-inch OD aluminum fuel line, using –6 tube nuts and ferrules. A five-way billet-aluminum fuel manifold from Peterson Fluid Systems (PN 10-0071) fed both carb's lines. This manifold has a single –10 male inlet and a staggered group of four –6 male outlets. The Peterson manifold also features two handy mounting tabs, so I could mount the unit to the rear of the tunnel ram intake via a fabricated bracket. I fabricated the mounting bracket that allowed attachment of the Peterson fuel manifold by using aluminum 2 x 1 L-channel stock. The two rearmost 1/4-inch x 20 cap screw locations secured the upper plenum mounts to the intake manifold. This location keeps the fuel lines away from higher-heat areas (as opposed to my original thought of mounting a fuel block to the valley cover).

Of course, cutting, bending, and flaring the tubing required a fair amount of hand bending to obtain a pleasing appearance. A hand tubing bender, a 37-degree flaring tool, and tons of patience were required as it took me three hours to complete. Once the lines had been fabricated, they were removed and I used a buffing wheel and polishing rouge to polish them to a satin luster. The final step was to hand-polish the lines with Wenol polishing paste. Lines were then washed, rinsed, and reinstalled.

Velocity Stacks

As if the induction system wasn't tall enough, I installed a pair of 8-inch-tall spun aluminum velocity stacks. A stamped aluminum screen disc rests on an inside-diameter lip of each stack. A 1/4-inch x 20 stud and nut secures the screen to the carb, which holds each stack securely in place.

Valve Covers

Because I chose to use the Taylor-Vertex valve cover adapters that allow mounting Generation I small-block Chevy perimeter-bolt

The fuel lines were painstakingly fabricated using 3/8-inch aluminum tubing and fed through a rear-mounted fuel distribution manifold.

valve covers to LS heads, I obviously had a choice of a wide range of valve cover styles. I chose a pair of relatively inexpensive cast-aluminum flattop-style covers from Summit Racing, PN SUM-G3302. Both covers each feature a 1.220-inch hole (one for a breather and one for PCV). Ace Powdercoating in Green, Ohio, did the powdercoating in a "silver splash" pebble-textured finish and Art Carlton at Innovators West engraved "408 *LS*" on each.

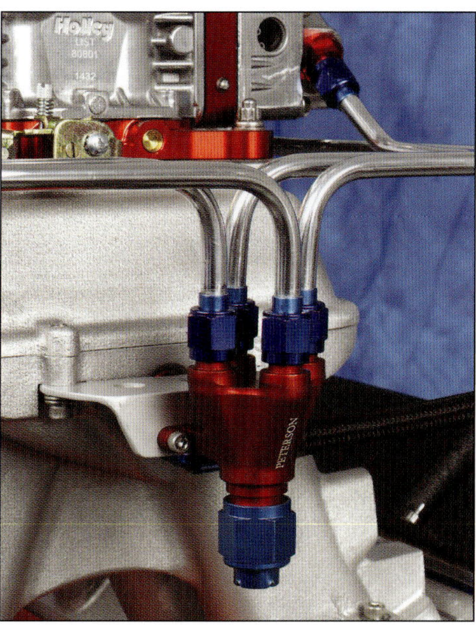

In order to feed our fuel line assembly, a fuel inlet manifold provided two feed lines to the front and rear of each double-pumper carb.

Ignition

Since we're running this engine with carbs, we don't need electronic management to control fuel, but we do need a system to control spark. The *easy* way to accomplish this is

While a more conservative air intake could have been employed, we decided to err on the side of ridiculous with a pair of polished aluminum velocity stacks, fitted with foam filters.

with MSD's LS ignition controller. It offers two versions: the 6LS, compatible with a 24-tooth crank reluctor, and the 6LS2, compatible with the 58-tooth reluctor. Our Callies crank came equipped with a 58-tooth wheel, so our ignition controller is the 6LS2.

The controller can be mounted anywhere in the engine bay within reach of the MSD harness. It features six preprogrammed ignition curve

Taylor/Vertiex Valve Cover Adapters

Spacer-Max-Inch PN 555701
(1999–2009 LSx)

The kit includes Fel-Pro silicone bead gaskets PN VS50504R (or you can use GM PN 12560696 gaskets). These bead gaskets create the seal between the cylinder head's valve cover rails and the adapters. For sealing the Chevy small-block valve covers to the Taylor adapters, Taylor recommends Fel-Pro cork-laminate gaskets, PN 1604 (5/16-inch thick).

The Taylor kit also includes eight 6-mm x 1.0 x 45-mm SHCS with lock washers and flat washers. Apply threadlocker and tighten the 6-mm SHCS to 106 in-lbs.

For attachment of the Chevy small-block valve covers to the adapter, the required thread size is 1/4 x 20 (length determined by thickness of the chosen valve covers). ■

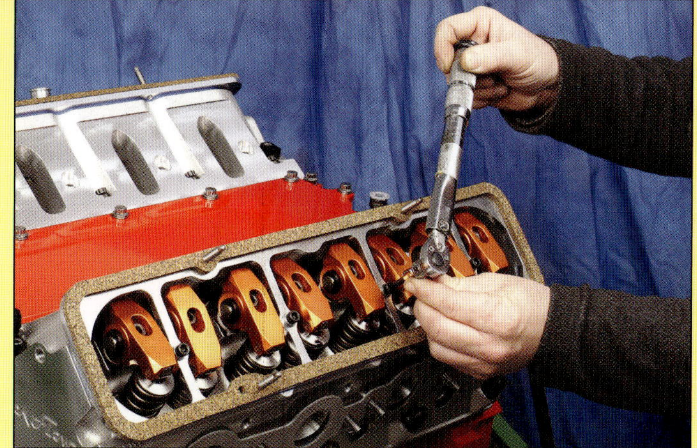

Aluminum valve cover adapters were secured to the heads in order to accept old-school small-block Chevy-style tall valve covers.

plug-in chips. This literally provides a plug-and-play selection. If you prefer, you can create your own custom ignition curve using the supplied CD (plug the controller into a laptop or desktop PC with a USB cable, and follow the instructions).

The MSD harness simply needs to connect to the ignition coils, the engine's water temperature sensor, cam position sensor, and crank position sensor (you can also connect to a MAP sensor). That's it. The ignition coils (just like any LS platform, you need one coil per cylinder) are MSD's PN 82858. These are listed as applicable to the LS1/LS6 but work just fine with a 58-tooth reluctor and the LS2-style cam sensor. No issues at all. NGK number-4177 spark plugs (3/4-inch reach, tapered seats) were installed with an adjusted gap of .054 inch. With anti-seize applied to the threads, all plugs were tightened to 10 ft-lbs. MSD 8.5-mm spark plug wires were obtained as PN 32079. This set features both straight "multi-angle" and 90-degree spark plug boots and terminals, and a set of LS-style coil boots and terminals (the multi-angle boots allow easy custom angle adjustment, from straight to almost 90-degree or anywhere in between). The multi-angle boots would protrude out between the primary tubes (with adequate clearance), while the use of 90-degree boots would allow me to route the wires behind the headers. After experimenting with each style, I decided to go with the multi-angle ends at the spark plugs.

Engine Oil

Brad Penn, Joe Gibbs, Royal Purple, and others provide break-in protection that's essential for flat-tappet as well as roller cam engines. For this application, I chose Brad Penn 20-50 break-in oil, high in zinc phosphate.

Our Dyno Results

A dyno session was conducted at Gressman Powersports, Fremont, Ohio, on its Superflow 902S dyno cell. The results: 670.5 hp at 6,500 rpm and 581.7 ft-lbs of torque at 4,900 rpm.

Aside from playing with timing, no issues arose. There were no leaks, no mechanical glitches, etc. We ran the ignition via an MSD 6LS2 ignition controller, which (when plugged into a laptop) allowed dyno shop owner Scott Gressman to quickly adjust timing during the session. While we were satisfied with the results, we're confident that additional testing time (messing with timing and carburetor tweaking) would likely provide us with an additional 10 to 15 hp, but we only had a specified amount of time allotted, since the shop had quite a few customers' race engines waiting their turn on the dyno. Listed here are the horsepower and torque results summary during the final pull for both torque and horsepower. Ignition timing was 9 degrees cranking, 15 degrees at idle, and (during incremental testing) was happiest at 26 degrees.

On the engine dyno, this 408-ci dual-carb build produced 670.5 hp at 6,500 rpm and 581.7 ft-lbs of torque at 4,900 rpm.

Block Specs

RPM	Ft-Lbs Torque	HP
3,700	484.5	367.2
3,800	495.3	383.9
3,900	507.8	402.2
4,000	520.7	421.3
4,100	529.0	437.4
4,200	539.5	455.5
4,300	558.1	473.2
4,400	559.5	491.1
4,500	567.7	509.3
4,600	574.6	525.8
4,700	579.4	540.6
4,800	581.5	553.2
4,900	**581.7**	564.0
5,000	580.8	573.9
5,100	579.5	583.3

RPM	Ft-Lbs Torque	HP
5,200	577.2	591.7
5,300	574.9	600.0
5,400	571.7	607.2
5,500	567.6	613.5
5,600	565.2	621.3
5,700	563.1	629.4
5,800	562.0	638.5
5,900	561.4	648.1
6,000	559.0	655.8
6,100	556.1	662.7
6,200	549.9	665.6
6,300	543.2	667.6
6,400	536.3	669.1
6,500	531.1	**670.5**

Dual-Carb LS 408 Parts List

Engine block	GM, 6.0L LQ9 Iron
Main caps	OEM PM (powdered metal) caps
Main studs	ARP, PN 234-5608
Main bearings	Mahle Clevite PN MS-2199HK
Rod bearings (upper)	Mahle Clevite PN CB-663HNK
Rod bearings (lower)	Mahle Clevite PN CB-633HXK
Cam bearings	Mahle Clevite PN SH-2125S
Cam bearing installer tool	Goodson PN CBT-300
Crankshaft	Callies Compstar, 4.000-inch stroke Compstar forged, 58-tooth reluctor, PN AP031N-CS
Connecting rods	Callies, 6.125 inches, Compstar forged, PN CSC6125 DS2A2AH
Pistons	JE flattop, FSR, Asymm, 4.030-inch bore, 1.115 CD, PN 311979
Cylinder heads	Trick Flow PN TFS-3061T001-C02
(Assembled heads actually used on dyno cylinder heads)	
Optional heads in bare form	Bill Mitchell Products, Warhawk, LS1, 235-cc int; 72-cc chambers, PN 001-025250
Threaded block plugs (x3) 16 mm x 1.5	GM PN 11588949
Threaded block plug, 28 mm x 1.25	GM PN 12561663
Front oil galley cup plug	GM PN 9427693
Rear oil galley plug barbell (15 mm x 64 mm)	GM PN 12573460
Cylinder head dowels (x4)	GM PN 12570326
Oil pressure sensor	GM PN 12616646
Water temperature sensor	GM PN 12608814
Crank position sensor (for 58-tooth)	GM PN 12585546
Carburetors (x2)	Holley 4150, 600 cfm, PN 0-80801-RD
Intake manifold	Holley Hi-Ram PN 300-226
Intake manifold top	Holley PN 300-216
Intake manifold bolts (to heads)	Totally Stainless, 8-mm x 1.25 x 80-mm SHCS
Upper plenum to intake manifold bolts	Totally Stainless, 1/4-inch x 20 x 3/4-inch SHCS
Carburetor linkage	Holley PN 4022
Carburetor velocity stacks	CP Performance PN 620-60445
Velocity stack stud kit	Allstar PN 26056 (studs cut to length)
Velocity stack stud nuts	Cam Motion Top Seal (1/4 x 20 threads)
Red 5/16-inch rod ends (for main throttle shaft)	QA1 PN AMR5
Vacuum fitting for PCV	Earl's PN 984206ERL
Valve cover breather	Summit Racing PN SUM-G3403
Valve cover breather grommet	Moroso PN 68772
PCV Valve grommet	Moroso PN 68775
Oil pump	Melling high-volume, PN 10296
Crankshaft damper	Fluidampr PN 740111
Crankshaft damper bolt (16 mm x 2.0)	ARP PN 234-2503
Cylinder head studs	ARP PN 234-4316
Exhaust headers	Hooker, black ceramic, PN 2292-3HKR
Header flange studs	Totally Stainless, 8 mm x 1.25 x 40 mm
Cylinder head gaskets	Mahle Victor, MLS, PN 54445
Gasket set	Mahle Victor PN CS5975
Engine pre-oiler/pressure unit	Goodson PN EPL-120
Camshaft (hyd. roller)	Comp PN 54-462-11
Pushrods (5/16-inch x 7.500-inch x .080-inch wall)	Comp Cams PN 7957-16
Lifters (short/race-hyd.)	Comp Cams PN 15850-16

Lifter guide trays	GM PN 12595365 (Gen IV) or 12551162 (Gen III) (Either style fits the LQ9 block)
Note: Drill outboard sides for better oil drain	
Intake valves (for BMP heads)	2.080 x 5.300 x 5/16 inch BMP PN 702830RM
Exhaust valves (for BMP heads)	1.600 x 5.300 x 5/16 inch BMP PN 702715SD
Valve spring kit (for BMP heads)	Comp Cams 26926TS-KIT
Rockers (rollers: 1.7:1 ratio)	Harland Sharp PN SHLS17
Oil pan	Holley, cast aluminum, PN 302-1
Oil pan bolts	ARP PN 434-6902
Oil filter	AC Delco PF48 or equivalent
Timing set	Cloyes PN 9-3172A
Cam gear bolts	ARP PN 134-1003
TFS Timing chain damper (LS2 style)	TFS-30675540
TFS Timing chain damper adapter	TFS-30675600
Camshaft retainer plate bolts	ARP PN 134-1002
Water pump	Meziere PN WP319U
Water pump waterneck	Meziere PN WN0019U
Water pump bolts (6)	Totally Stainless, 8 mm x 1.25 x 115 mm
Front cover (OEM)	GM, LS2 (with cam sensor) GM PN 12633906
Front cover (Comp)	Comp Cams PN 5496
Front cover bolts	ARP PN 434-1502
Rear cover	GM PN 12633579 (PN 12639250 KIT)
Rear cover bolts	ARP PN 434-1504
Valley cover	Birchwood, fabricated (.250-thick aluminum)
Valley cover bolts (x12)	Totally Stainless, 8 mm x 1.25 x 20 mm
Engine block paint	Chevy Orange-Red (Dupont base/clear)
Spark plugs	NGK 4177 or equivalent
Valve cover adapters	Taylor/Vertex PN 555701
Valve covers	Summit PN SUM-G3302
PCV valve	M/E Wagner, billet adj., PN DF-17
PCV valve hose	–6 black braided (with shrink tube ends)
Oil cooler adapter kit	Lingenfelter PN L300025297
Steam hole front fittings (-4 AN)	TFS-30600611
Steam hole rear caps	TFS-30600612
Front steam plumbing	Two –4 90-deg. hose ends; two –4 straight hose ends, –4 T-fitting, –4 black braided hose
Fuel lines	3/8-inch OD aluminum, fabricated
Fuel line fittings	–6 tube nuts and ferrules
Fuel manifold	Peterson PN 10-0071 (five-way –6AN/–10AN)
Fuel manifold bracket	Birchwood (fabricated aluminum)
37-degree flaring tool	Fragola PN 900500
Ignition controller	MSD PN 6LS-2 (for 58-tooth reluctor)
Ignition coils	MSD PN 82858
Spark plug wires	MSD PN 32079
Dipstick assembly	Lokar PN 5008
Timing pointer	TCI PN 871005

CHAPTER 11

DART LS NEXT BUILD

The Dart LS Next is an aftermarket iron block that's the basis of this LS build. While it incorporates the architecture of GM's LSX race block, Dart took it a step further and instituted a host of upgrades, including priority main oiling, thicker decks, stronger cylinder walls, improved cooling passages, reinforced main webs, and elimination of GM's Y-block design by eliminating 2 inches from the pan rail areas for improved bay-to-bay breathing and reduced windage concerns. This block has enormous strength for racing and extreme supercharged and turbocharged applications. With this particular project, we opted for a 4.185-inch bore and 4.000-inch stroke for a displacement of 440 ci. Compression ratio is a pump-gas-friendly 10.54:1. Equipped with Trick Flow LS3-style CNC heads and JE inverted dome pistons, this build pulled 665.2 hp at 6,000 rpm and 627.3 ft-lb of torque at 4,900 rpm.

The LS Next Block

This block has many important upgrades that eradicate several weaknesses and compromises of the factory blocks, producing a notably stronger bottom end for those who chase serious grunt. Pan rails are factory width but have been shortened, removing the Y-block design. You need to fit a special oil pan, such as a Stef's, Canton, or Moroso rail adapters, to the LS Next block. Eliminating the LS engine's Y-block design addresses the problems from the OEM LS engine's separated crankcase bays. Windage has been significantly lowered, which improves power production and it also has a stronger main web design.

Thread sizes on the Dart LS Next block retain all of the metric sizes featured on the factory block, with the

This 440-ci build began with a Dart LS Next block featuring Trick Flow LS3 style heads. Displacement was achieved with a 4.185-inch bore and 4.000-inch stroke.

Dart's LS Next block is available in either cast iron or aluminum. This block is vastly superior to the OEM factory block in terms of strength, windage reduction, oil delivery, and versatility. The most noticeable feature is the elimination of the Y-block extended pan rail design. This improves crankcase breathing and eliminates the need for the 8-mm main cap pinch bolts.

exception of cylinder bolt and main cap holes, which are 7/16-14, the outboard cylinder head bolt holes, which are 3/8-16, and the array of NPT plugs (1/4 inch and 1/8 inch NPT). It requires a remotely plumbed oil filter.

All remaining threaded holes in the LS Next block retain the OEM-size metric threads. Actual deck height is .002 to .005 inch taller for additional machining requirements.

Upgraded features include thicker decks, priority main oiling, and improved water flow and capacity. Versions are available with either standard camshaft tunnel location or raised-cam designs to provide added stroke clearance. The blocks will also accommodate 23-bolt heads. In addition, 1/4-inch NPT ports are featured for solid lifter applications. Since we ran hydraulic roller lifters, we plugged these ports.

The LS Next block is fitted with four-bolt billet-steel main caps for added bottom-end strength.

According to Dart, you need to install the factory-style Cloyes timing gear PN 9-3658TX3. Double-row timing chains require block clearancing. Head stud holes are 7/16 inch x 14, are blind, and do enter water jackets.

When using cylinder head studs, you should use a bull-nosed stud design so the unthreaded portion of the stud does not dig into the female threads. With a flat-tipped stud design, place a small bearing ball into the 7/16-inch blind hole, so the unthreaded shoulder of the stud cannot be forced against the top of the threaded hole. If using flat-tipped studs, each stud needs to be test fitted so you can arrive at the correct ball diameter (1/16, 3/32, or 1/8 inch) to raise the stud accordingly. In the end, the tapered portion of head studs must not contact the deck surface.

The LS Next block uses 1.5-inch press-in freeze plugs, rather than LS threaded plugs. The block also comes with the camshaft retaining plate and unique bolts (PN 32226000). The left side of the block is machined for –10AN straight thread O-ring fittings. The filter adapter for the required remote oil filter connects with –10 AN hose and hose ends.

Priority Main Oil System

The oil-in hole is found above the pan rail on the left-side rear while the oil-out is situated above the pan rail on the block's front left side. Oil routes to the main bearings and then to cam bearings. The main oil galley sends oil to the crossovers and then to the lifters. If lifter oiling needs to be restricted, install 1/8-inch NPT plug restrictors in the center crossovers just above the main oil galley. Dart recommends starting with a restriction of .051 inch with modified lifters. Blocks are machined, so they can be fitted with six-bolt-per-cylinder heads (known as 23-bolt heads). In addition, you can upgrade the 7/16-inch head bolt locations to 1/2-inch fasteners for greater clamping force.

Cylinder Head Decks

The main deck uses 10 7/16-14 threaded holes. Five 8-mm x 1.25 threaded holes are located at the inboard edge of the deck. Four 3/8-16 threaded holes are located at the outboard edge of the deck. Four .400-inch holes are found in the area between the cylinder bores and the lifter well, and these are designed to accommodate the four "extra" cylinder head studs on a 23-bolt head. Our ARP 23-bolt head stud kit has 8-mm x 1.25 x 69-mm studs that mount to the deck of the head before positioning the head onto the block deck. Special shouldered washers and nuts are accessed from the lifter valley.

Machining the LS Next Block

Gressman Powersports in Fremont, Ohio, completed the finish machining of the block. Scott Gressman used his CNC block machining center to precisely bore, deck, and square the block. He initially measured the cylinders at 4.125 inches. He performed the boring process in two steps. First, he bored to 4.155 inches, and then 4.180 inches, so .005 inch was left for final honing. To start with, block deck height measured in a range of 9.2375 to 9.2450 inches. We need to square the block decks, so the crank centerline needed to be equal distance from all deck surfaces. When the machining was completed, deck height was finished at 9.2370 inches. The CNC machine was then used to slightly chamfer all top edges of each cylinder bore.

Using the same ARP head studs for final assembly, the deck plates were torqued to the block. Gressman started the honing process using the Rottler 275 to 325 diamond stones to size. He took four strokes using 500-grit stones and then he took four strokes with plateau brushes to complete the finish hone. JE states that piston skirt diameter needs to be measured at a point .275 inch from the bottom of the skirt. Our skirts measured 4.1795 inches. The finished diameter resided at the JE specified .0055-inch piston-to-wall clearance. Gressman also verified size and alignment of the main bores.

LS Next Block

Fastener	Dimensions
Head bolt threads in block decks	7/16-14, 3/8-16, 8 mm x 1.25
Main cap bolts	7/16-14
Bellhousing bolt holes	10 mm x 1.5
Oil pan rails	1/4-20
Cam retainer plate	6 mm x 1.00
(Button-head screws supplied with block require 4-mm hex wrench. Block does not accept OEM retainer plate; Dart's plate must be used.)	
Oil pump to block	8 mm x 1.25
Front cover to block	8 mm x 1.25
Rear cover to block	8 mm x 1.25
Valley cover to block	8 mm x 1.25
Crank sensor to block	6 mm x 1.00
Engine mount bolt holes	10 mm x 1.5
Water pump bolt holes	8 mm x 1.25
Lifter tray mount bolt holes	6 mm x 1.00 (with OEM lifter trays)
Front of block oil galley holes	1/4 NPT
Rear of block oil galley holes	1/4 NPT
Left- and right-side water drain holes	1/4 NPT
Left side of block for remote filter	–10 AN straight with O-rings
Oil restrictor holes in valley	1/4 NPT
Oil holes in left center lifter wells	1/8 NPT
Expansion plug holes, block sides	1.5 inches
Expansion plug holes, block rear	1.5 inches
Note: The Dart block completion kit (available as Dart PN 32000016) includes six 1.5-inch freeze plugs, two bellhousing dowels, four head deck dowel sleeves, two 1/4 NPT male hex head plugs for water drains, seven 1/4 NPT female hex plugs, and three 1/8 NPT female hex plugs.	

Threaded Block Plugs/Fittings		
All NPT plugs were installed with a coating of Teflon pipe paste on the threads.		
Qty	**Size**	**Location**
3	1/4 NPT	Front of block, above cam bore
2	1/4 NPT	Rear of block
1	3/8 NPT	Rear of block
2	1/8 NPT	Valley (oil restrictor intersecting holes)
2	1/4 NPT	Valley (oil restrictor access holes)
2	1/4 NPT	Each side of block/water drain holes
1	1/8 NPT	Oil pressure hole, left lower rear of block
2	–10 AN fittings with –10 AN straight thread fittings with O-rings for plumbing to remote oil filter	

Crankshaft Balancing

Medina Mountain Motors in Creston, Ohio, precision balanced the Scat forged crankshaft. First, pistons, rod big and small ends, rod bearings, rings, pins, and pin locks were weighed, and then bobweights were installed on the crank. The crank was internally balanced to within .2 gram. Heavy metal addition was not required, so removing weight from the front and rear counterweights achieved the desired balance. A total of 18.125 grams was removed.

Our Bobweight Card

Piston	456 g
Pin	115 g
Rings	41 g
Locks	2 g
Rod small-end	187 g
Rod big-end	431 g
Rod big-end	431 g
Rod bearing	86 g (43 x 2)
Total	1749 g
1/2 total	874.5 g
+ 4 g oil	878.5 g

Crankshaft Installation

During test fitting, I first checked counterweight-to-block clearance. Dart obviously had stroker cranks in mind when they designed the block. I found no clearance issues whatsoever. The closest counterweight-to-block areas were perhaps .300 inch or so.

Main bearing clearance measured .002 inch. Crank endplay was measured at .005 inch.

During final installation, the 7/16-inch main cap bolts were snugged to 65 ft-lbs with CMD

During crank test fitting, main bore bearing oil clearance checked at .002 inch with rod bearing clearance at .002 inch. Crank thrust was verified at .005 inch.

The LS Next build used JE forged pistons that featured an inverted dome with volume at –14.6 cc. Our final bore diameter was 4.185 inches with piston-to-wall clearance of .0055 inch.

number-3 high-pressure lubricant, which is the same torque used during main bore alignment check.

I tightened the bolts in stages, starting at 30 ft-lbs, then 45 ft-lbs, then a final 65 ft-lbs. Crank rotation was checked after each stage, with the crank turning like butter each time (I could rotate the crank with two fingers on the snout). The Mahle Clevite main bearings were coated with Royal Purple Max Tuff assembly lube.

Piston Specs

PN 324072	
Inverted dome	14.6 cc
Inverted dome depth	.097 inch
Comp. distance	1.115 inches
Bore	4.185 inches
Valve pockets	12 degrees

Site	Groove Width	Land Widths
Top	.048 inch (1.22 mm)	.255 inch (6.48 mm)
Second	.060 (1.52 mm)	.165 inch (4.19 mm)
Oil	.149 inch (3.78 mm)	.090 inch (2.29 mm)

Radial Depths		Pins	
Top	.168 inch (4.27 mm)	Pin diameter	.927 inch
Second	.177 inch (4.50 mm)	Pin length	2.250 inches
Oil	.172 inch (4.37 mm)	Wire locks	.073 inch

Our Measured Clearances

Piston to wall	.0055 inch
(Skirt diameter measured .275 inch from bottom of skirt)	
Main bearing clearance	.0017 inch
Rod bearing clearance	.0020 inch
Crank thrust clearance	.0065 inch
Rod sideplay	.018 to 0.019 inch
Lifter-to-lifter bore clearance	.0015 inch

Pistons

The design of our JE pistons is asymmetric. The major load side features a conventional-width skirt, while the unloaded side skirt is narrower. This saves weight and provides the thrust side of the piston the support needed while the non-thrust side does not require a large skirt because it's there simply to provide piston stability, and is able to use a narrower skirt area. The design goes beyond this difference in skirt surface area. The piston is designed internally to provide maximum strength for the loaded thrust side while relieving weight where mass can safely be removed. The on-center balance weight of the piston, relative to the pin bore centerline, is altered due to design, so the centerline of the wrist pin bore was moved about .020 inch toward the larger skirt side to compensate, achieving a centered weight at each side of the pin bore axis.

Connecting Rods

Our Scat forged 4340 steel H-beam rods, PN 2-350-6125-2100-S, are the firm's LS rods featuring high-strength H-beam design, doweled caps for cap-to-rod register, and a profile clearanced for stroker applications. The rods feature a center-to-center length of 6.125 inches. The rods are fitted with ARP 8740 12-point rod bolts, featuring 7/16-inch shank diameter and a shank length of 1.400 inches.

Installing the rods and pistons was aided with my adjustable Summit Racing piston ring compressor. A split design allows adjusting for bore diameters of 4.120 to 4.220 inches, with a tapered inner wall that smoothly compresses the ring package as the piston slides through the compressor. Once compressor diameter was finely adjusted to 4.185 inches, I was able to push each piston into its bore smoothly by hand, with a closed fist. Once the piston was engaged into its bore, I used Goodson's extended-snout plastic piston hammer to gently tap the piston down while aligning the big end onto the crank journal.

Rod-bolt-to-block clearance was previously checked during test fitting. Dart had already notched the bottom of the bores for rod bolt clearance. Our closest rod-bolt-head-to-block spacing was approximately .300 inch, so no additional block notching was required.

Scat specifies tightening torque at 64 ft-lbs with ARP moly, not to exceed .0046-inch bolt stretch. I first zeroed my rod bolt stretch on each rod bolt on gauge in its relaxed state. Once each bolt was torqued to 64 ft-lbs, rod bolt stretch was measured at .0035 to .004 inch. To make the process quicker, I used two stretch gauges, one for each rod cap's bolts.

During test fitting of the rods, I determined that my desired .002-inch oil clearance at the rod bearings would be achieved by running Clevite HNK rod bearing in the upper locations, along with HXNK lower bearings. With rods installed to the crank, rod sideplay was measured at .0018 to .0019 inch at all rod big-end locations.

Camshaft

For this build, I chose Comp Cams' hydraulic roller, PN 54-462-11. Camshaft endplay was measured at .006 inch. Comp Cams recommends an endplay of .005 to .010 inch.

Timing Set

After installing a key in the crank snout's rear (long) keyway, the Cloyes crank gear was tapped into place with an interference fit of approximately .0015 inch. The "standard" timing slot was engaged to the key (the rounded slot). This placed the zero

Camshaft Specifications

Gross valve lift (with 1.7:1 ratio rockers)	.624 intake/exhaust	
Lobe lift	.3670 intake/exhaust	
Duration at .006-inch tappet lift	292 intake/300 exhaust	
Valve timing at .006	Intake open	34 BTDC Close 78 ABDC
	Exhaust open	86 BBDC Close 34 ATDC
(Installed at 111.0 intake center line)		
Duration at .050	243 intake/251 exhaust	
Lobe separation	114.0 degrees	

Compression Ratio

Number of cylinders	8
Cylinder bore	4.185 inches
Stroke	4.000 inches
Rod center to center	6.125 inches
Gasket bore diameter	4.200 inches
Compressed gasket thickness	.051 inch
Block deck height	9.2370 inches
Combined chamber volume	69 cc
Piston dome volume	+14.6 cc
Piston to deck	.003 inch
Total volume	996.19 cc
Cylinder volume	901.69 cc
Clearance volume	94.5 cc
Gasket volume	11.57 cc
Deck volume	– 0.68 cc
Piston top land	4.147 cc
1/2 stroke	2.000 inches
Piston compression height	1.115 inches
Displacement	440.18 ci
Static compression ratio	10.54:1

timing mark counterclockwise relative to the key. The Cloyes oil pump drive gear was installed onto the same key, with a slight interference fit of about .0015 inch.

Important: The black side of the cam gear's Torrington bearing must face the engine block. Install the eccentric cam timing button onto the drive pin hole in the cam gear. The rear of the button engages onto the camshaft's dowel pin. The eccentric button can be adjusted with a 1/4-inch hex bit at zero or advance/retard, to allow you to fine-tune cam timing during cam degreeing.

After soaking the timing chain in 30W engine oil overnight, I installed the cam gear and chain (with bearing and adjuster button), tightening the ARP cam gear bolts to 25 ft-lbs (with Loctite 242).

Piston/Rod Assembly

During assembly, we need to keep in mind that the pistons are asymmetric and dedicated for right and left banks. To ease organizing and orientation, each piston dome is lightly etched by JE with a "FRONT" designation and an arrow symbol. When installing pistons to rods, also keep in mind that the larger chamfer on one side of each rod big end must face forward on the left bank and rearward on the right bank.

The JE pistons feature full-floating pins, secured at each end by a single round wire clip. Installing wire clips is a bit of a nuisance, but you simply need to deal with it. One trick that Jody of Medina Mountain Motors revealed is to turn the clip 90 degrees to the pin bore, entering the bore with the side of the clip opposite the clip gap. Using both thumbs, push the clip into the bore, just past the clip groove. Then slide a piston pin through the opposite side of the piston, contacting the clip. Continue to push the pin until it straightens out the clip and pushes it toward the groove, until the clip snaps into the groove.

For the opposite side, insert one end of a clip into the piston pin bore groove, about 90 degrees away from the notch at the edge of the pin bore.

Piston Rings

Our JE set included all rings, plus a set of support rails that install onto the bottom of the oil ring land, providing oil ring support at the outside of each pin bore end (necessary because the compression height of our pistons requires the pin bore to slightly intersect the oil ring groove). Install each support rail so that the rail's small male bump aligns at the center of the pin bore area. This small bump serves as a stopper to prevent the rail from walking, keeping the rail's gap well away from the open pin bore cutouts. The side of the support rail that features the male bump must face downward.

Our ring set includes top rings to accommodate 1.22-mm groove width, second rings to accommodate 1.52-mm groove width, and oil ring packages to suit 3.78-mm oil rings and support rails. Both second and top rings required file fitting for our 4.185-inch bore size (a common procedure for any modified bore diameter).

JE recommends ring end gaps of bore size x .0045 for top rings and bore size x .0050 for second rings. Using my Summit Racing bench-mounted ring filer, I obtained a .019-inch gap for all top rings and .021-inch gap for all second rings, followed by carefully deburring all filed edges.

I file fit each top and second ring on a per-cylinder basis.

Oil Pump

Per Dick Maskin at Dart, I swapped out the original red pressure spring from the Melling 10295 oil pump and installed a standard "green" spring from a Melling M55 small-block Chevy pump. Dick advised that the LS Next block's oiling system, a product of extensive research and development in-house at Dart, is so much more efficient than the OEM design that a lower-pressure spring is needed in an LS pump. Where the original red spring is rated at 60 to 70 pounds, the M55 spring is rated at 49 pounds. Dick noted that during development work, the excessive oil pressure created by the higher-pressure spring blew out an oil filter.

Engine Covers

For this build I selected a new stock GM rear cover, but for the front cover I chose Comp Cams' LS cover, which is deeper to provide added clearance and features a provision for easy mounting of a timing pointer (your choice left or right side). A billet spacing adapter is included to allow mounting a timing pointer that's part numbered for a big-block Chevy application. Also, the provision for the front-mount cam position sensor is unique. A block-off plug is included in case you are using a block with a rear-mount cam sensor, and an insert is provided that accepts the front-mount sensor. If the block features a standard cam location, install the fitting with the hole biased toward the crank, at six o'clock. If you have a raised-cam block, orient the fitting with the sensor hole up, at twelve o'clock. The fitting features a groove and a square-cut O-ring to seal the fitting to the cover. The Comp cover also features threaded stanchions that allow mounting a big-block Chevy timing pointer and/or a crank trigger ignition pickup (two threaded holes per side are featured).

I installed a TCI timing pointer. Comp designed the cover to accept a readily available big-block Chevy timing pointer. Comp's front cover kit includes a machined aluminum adapter to permit pointer installation. The adapter mounts to the cover with a pair of 1/4 x 20 x 3/8 socket head cap screws. The TCI pointer then attaches to the adapter with a pair of 1/4 x 20 stainless-steel socket head cap screws.

I had a valley cover fabricated at a local CNC shop, using a piece of flat-stock 6061 aluminum, .250-inch-thick x 6 inches x 20.375 inches. Features include three ball-milled grooves on top, 45-degree chamfered outer top edges, and a CNC-milled O-ring groove on the underside to accept a strip of 3/32-inch butyl rubber oil-resistant O-ring cord.

Once the front and rear covers were powdercoated, I installed new Victor seals from our gasket kit by running the mounting bolts just a few threads past the gaskets (the gasket bolt holes feature bead-formed seals that grip the bolt threads, making it easier to keep the gaskets aligned with the covers).

The camshaft position sensor wire harness was removed from its factory bracket, and the wires were shrouded with a black braided wire sheath. To secure the cam sensor wire harness, I made a small aluminum bracket that bolts to the left side of the front cover with a pair of 3/8 x 16 x 1/2 bolts (using provided threaded holes in the cover). A small Adel clamp attaches to the bracket with a single number-8 x 32 x 1/2 stainless socket head cap screw and nyloc nut to prevent the lower section of the harness from flopping around.

Oil Pan

The Moroso oil pan kit has been specifically designed for the Dart LS Next block. The welded steel oil pan features a pair of billet-aluminum pan rail spacers and a windage tray. The side spacer bars establish a flush pan and gasket surface in conjunction with the bases of the OEM-style front and rear engine covers.

The billet spacers attach to the block's side rails with 1/4-20 to 1/4-28 studs (five per side). The oil pan then attaches to the billet rails with a series of 5/16-18 to 5/16-24 studs. Strips of Moroso-supplied O-ring material provide sealing between the block and the billet rail spacers.

The underside of the front cover secures to the oil pan with a pair of 8-mm x 1.25 studs and nuts, and the underside of the rear cover secures to the oil pan with a pair of 6-mm x 1.0 studs and nuts. Once the spacer rails are in place, the oil pan mounts with 5/16-inch studs along the sides. Two 8-mm studs are installed at the bottom of the front cover and two 6-mm studs are located at the bottom of the rear cover.

Moroso offers an oil pump pickup that's specially designed for the LS Next block and the Moroso oil pan. The pickup secures to the oil pump inlet port (typical LS design), and the pickup tube secures to one of the main cap studs.

Note: Neither the block nor the Moroso oil pan has a provision for a dipstick. This simply requires drilling an angled hole on the passenger's side of the sump wall and welding in a .710-inch OD steel tube that features a 1/4 NPT female thread at the outer end.

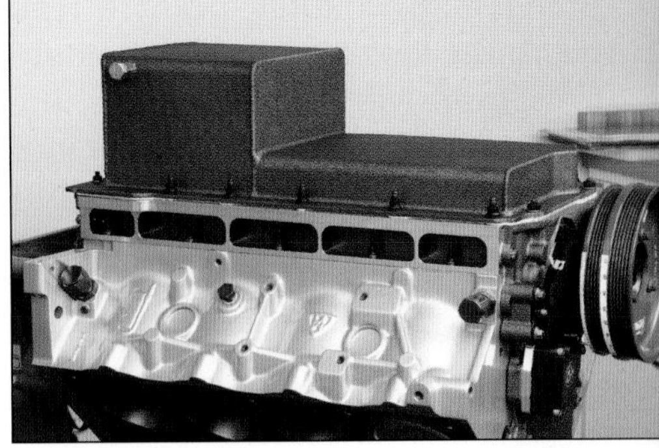

The 2-inch-tall aluminum spacer rails included with the Moroso oil pan kit are specially designed to accommodate the unique LS Next installation. The cutouts you see are milled to provide access for the mounting bolts. We simply painted the cutouts black to provide added visual appeal.

When installing the oil pan kit, first install the O-ring strips to the grooves in the top surfaces of the billet rails. Make sure that the O-rings are fully seated.

Install the 1/4-20 stud bottoms to the block rails (with a dab of Loctite 242), finger tight.

Install the billet rails to the block, tightening the 14 to 28 nuts to 18 ft-lbs. Because of the captive location inside the billet rail pockets, you'll need a torque wrench extension (if the extension is 2 inches long, you need to reduce the setting on the torque wrench to compensate for the extension). For the 12-point 1/4-28 nuts, you'll need a 5/16-inch torque wrench adapter/extension. An example is Snap-On's PN FRDH101 (5/16-inch 12-point box at one end and 3/8-inch square drive at the other end).

Adjusting the Torque Wrench When Using an Extension

Use this formula to determine the setting:

$$[L / (L + E)] \times TE = TW$$

L = length of torque wrench (grip center to ratchet head)
E = length of wrench extension
TE = actual applied torque/desired torque
TW = torque wrench setting

If the torque wrench is 14 inches and the wrench extension is 2 inches: L = 14 and E = 2

$$14 / (14 + 2)$$
$$14 / 16 = .875$$

So, .875 x desired torque = wrench setting

The 1/4-24 billet-rail-to-block nuts are specified for 18 ft-lbs. Example: if using a 14-inch-long torque wrench and a 2-inch-long wrench extension, set the torque wrench at 15.75 ft-lbs. ■

Loosely install the front and rear covers with 8-mm x 1.25 bolts, along with OEM-type cover gaskets.

Install the 5/16-inch studs to the bottom of the billet rails (with a dab of Loctite 242, finger-tight).

Install the 8-mm x 1.25 studs (with Loctite 242) to the bottom of the front cover and install the 6-mm x 1.0 studs (with Loctite 242) to the bottom of the rear cover. These front and rear cover bottom studs feature a female hex and allow you to adjust the installed depth.

Apply a dab of RTV to each corner (where rails meet front and rear covers).

Install the OEM-style oil pan gasket.

Install the oil pan. Snug all the pan fasteners to allow the front and rear covers to mate flush with the pan (tightening the front-pan-to-front-cover bottom 8-mm x 1.25 stud nuts and the 5/16-24 stud nuts to about 10 ft-lbs and the rear-cover-to-pan 6-mm x 1.0 stud nuts to about 60 in-lbs).

Note: The front and rear ends of the billet rails feature open holes that align to the lower mounting holes in the front and rear covers. Install a 5/16-inch bolt through the cover and into the rail, and secure with a locking nut. Access to the rear horizontal bolt nuts is tight due to the proximity of the rear 1/4-inch rail mounting studs.

Tighten the front and rear cover bolts to 18 ft-lbs.

Final-tighten the oil pan stud nuts (front-cover-to-pan 8-mm nuts to 18 ft-lbs; 5/16-inch stud nuts to 25 ft-lbs; and the rear-cover-to-pan 6-mm stud nuts to 8 ft-lbs or 96 in-lbs). To keep the desired stud installed depth of the front 8-mm studs and the rear 6-mm studs, you may wish to hold these studs stationary with a hex bit while torquing the nuts, although the Loctite should prevent them from moving.

Lifters

For this build, we opted for a set of top-of-the-line super-strength, close-tolerance/high-precision Morel hydraulic rollers, featuring a V-style tie-bar design (PN MP5294). As opposed to using individual lifters that require a separate lifter guide, these lifters drop in as a pair, with the tie bar serving to maintain lifter orientation (roller to cam lobe). The V design of the tie bar provides the needed clearance of the "extra" head studs on our 23-bolt block and head design.

Morel link-bar lifters are fitted with V-bar links. In some cases, the bent V-bars provide added clearance and access to the extra studs involved with the LS3-style heads.

Keep in mind that the Dart LS Next block is designed for the six-bolt head pattern. This features a drilled boss that protrudes into the center of each lifter well area. If you plan to use tie-bar lifters, the tie bar should feature a "V" bend that clears the extra 4-cylinder head studs at each bank.

To check lifter bore oil clearance, I first measured lifter diameter at .841 inch with a micrometer.

Using Goodson's FBG-7 dial bore gauge (range of .700 to 1.50 inches), I indexed the gauge to match .841 inch and zeroed the gauge needle. I then used the bore gauge to check each lifter bore, measuring a clearance of .0015 inch on each bore, with zero taper and zero out of round. Considering lifter diameters and measured clearance, our lifter bore diameters are .8425 inch.

Our Morel hydraulic roller lifters feature a roller diameter of .750 inch. Morel's John Callies recommends lifter-to-bore clearance in a cast-iron block at .0015 to .0017 inch. After achieving zero lash, preload (cast-iron block and aluminum heads) is recommended at .030 to .035 inch.

Callies recommends coating each lifter with 10W40 oil for installation (do not wash the lifters in any type of solvent prior to oiling). For final engine operation, he suggests 0W40 or 5W40 oil. He prefers a full synthetic, such as Mobil 1. Due to the tight clearances involved in lifter design, this viscosity range is appropriate.

Timing chain damper/rub blocks are available from GM or aftermarket suppliers. Shown here is a tan-colored GM block.

Water Pump

I often install Meziere electric water pumps on my engine builds because they are ultra reliable, perform flawlessly, and have an attractive appearance. Our blue anodized billet-aluminum pump (WP319B) was installed with six 8-mm x 1.25 x 80-mm stainless-steel socket head cap screws, torqued to 22 ft-lbs. This pump model features a high-flow idler for serpentine belt applications and is flow rated at a stout 55 gpm.

You need to take the following into account. The driver's side of the Meziere water pump main body edge contacts the front cover's sensor boss when installing it with a Comp Cams front cover. In the past, I've faced this problem when using a combination of a Meziere pump and Comp front cover. This isn't a design problem of either component. The Mezsiere pump is designed to clear

Crank Damper and Cylinder Head Specs

Torque Specifications	
Main cap bolts	65 ft-lbs with cmd #3 high-pressure lube
ARP head stud nuts 7/16 inch	75 ft-lbs with cmd #3 (*note procedure!)
ARP head stud nuts 3/8 inch	35 ft-lbs with cmd #3
ARP head stud nuts 8 mm	22 ft-lbs with cmd #3
ARP oil-pump-to-block bolts	18 ft-lbs
Cam retainer plate screws	86 in-lbs
ARP cam gear bolts	25 ft-lbs with Loctite
Timing chain damper to block	18 ft-lbs
ARP crank damper bolt	235 ft-lbs
Oil-pickup-to-oil-pump bolt	106 in-lbs
Oil pickup support bracket nut	35 ft-lbs
Front cover bolts	18 ft-lbs
Rear cover bolts	18 ft-lbs
Rod bolts	64 ft-lbs with moly (not to exceed .0046-inch stretch)
Oil pan rails to block	18 ft-lbs
Oil pan to rails (5/16 stud nuts)	25 ft-lbs
Oil pan to front cover (8-mm stud nuts)	18 ft-lbs
Oil pan to rear cover (6-mm stud nuts)	96 in-lbs
Meziere water pump bolts	22 ft-lbs

Cylinder Heads
Trick Flow GenX, PN TFS-3261T004-C01
255-cc intake volume
CNC competition ported
69-cc CNC combustion chambers
448-pound springs
Titanium retainers
Six-bolt
LS3-style with rectangular intake ports

Combustion chambers feature a volume of 69 cc.

an OEM front cover, but the Comp cover is designed to work with an OEM pump and is a bit thicker in this area. To resolve this issue, you lightly chamfer either edge of the boss on the cover or the contact edge on the lower left of the pump. If you are not aware of this, you'll wonder why the pump won't seat flush to the block when you're trying to install it.

Note: If you install an electric water pump, you should avoid routing the wire harness directly against the cam sensor harness, so any frequency issues for the cam sensor are negated.

Crank Damper

Our 7.5-inch Fluidampr is an integrally fluid-dampened unit featuring a built-in grooved pulley. I applied a light film of anti-seize to the damper bore and I installed the damper with the ARP crank bolt and torqued it to 235 ft-lbs with ARP moly. It also has an interference fit of about .0016 inch on the crankshaft snout. I smoothly drew the damper onto the crank snout using a fabricated installation tool comprising a length of 16-mm x 2.0 all-thread and a steel driver that contacts the front recess (washer face) of the damper. The outboard end of the all-thread features two flats that allow the threaded rod to be held stationary while a nut (with lubed washer) is tightened against the steel driver.

To obtain a timing reference, MSD timing tape was applied to the center smooth surface between the front and rear belt ribs, adjusting our TCI timing pointer to zero at TDC with the number-1 piston at TDC on compression stroke. This was done prior to cylinder head installation.

Cylinder Head Stud Kit

ARP offers a cylinder head stud kit specific to the 23-bolt LS Next application. Each head utilizes the following studs:

Install *all* studs only finger-tight. Dart suggests applying Loctite 241 to the course threads, although this isn't necessary (but not a bad idea for the five 8-mm studs that screw into the cylinder head deck).

Qty	Lower	Upper	O/A	Wrench	Location
4	3/8-16	3/8-24	2¼-inches	1/2 inch	Outboard
5	8 x 1.25 mm	5/16-24	65 mm	3/8 inch	Inboard
4	8 x 1.25 mm	5/16-24	70 mm	3/8 inch	Extra*
2	7/16-14	7/16-20	3¾-inches	5/8 inch	Lower corners
8	7/16-14	7/16-20	3½-inches	5/8 inch	Primary

*The four (per head) "extra" 8-mm x 1.25 x 70-mm studs thread into the bottom of the head and pass through bosses in the block, secured with 5/16 x 24 nuts and special shouldered flange washers. A 3-mm hex wrench is required to drive (install/remove) these studs.

Aftermarket performance crankshaft damper/pulley assemblies are available to replace the factory pulley. In addition to providing a pulley surface for the serpentine belt, viscous-filled dampers (also called crank balancers) serve to dampen any harmonic crankshaft vibrations. The balancer shown here is from Fluidampr. This style of balancer is also much easier to install and remove as opposed to the factory cast pulley, allowing the use of a commonly available balancer installation/removal tool.

Cylinder Head Installation

Dart recommends using only CMD number-3 high-pressure lubricant for nut-to-stud lubrication. In addition, you should thoroughly apply muriatic acid to the upper fine threads of the 7/16-inch studs to strip the black coating. As a result of their extensive research, the personnel at Dart feel the coating has a variable effect on torque value. The 8-mm studs need to be mounted to the underside of the head, and then the head gasket positioned onto the block deck. Then correctly orient the gaskets on the head. The Fel-Pro MLS gaskets are clearly labeled "FRONT." Position each gasket so that the "FRONT" label is facing upward and is visible, with this end facing the front of the block.

The Trick Flow heads feature LS3-style rectangular intake ports with 255-cc volume.

Cylinder Head Torque

Dart specifies torque values for head installation to its block as follows:

Fastener Thread Size	Torque Value (Ft-Lbs)
7/16 inch	75
3/8 inch	35
8 mm	22

Per Dick Maskin's advice, I followed a very specific torque procedure for the 7/16-inch nuts:

1. Torque 7/16-inch nuts to 40 ft-lbs, then 50 ft-lbs, then 75 ft-lbs (in proper sequence).
2. Wait five minutes.
3. Crack all nuts loose (to zero ft-lbs).
4. Torque all 7/16-inch nuts directly to 75 ft-lbs (in same sequence pattern).
5. Crack nuts loose again to zero.
6. Final-torque 7/16-inch nuts directly to 75 ft-lbs.

Note: I loosened all 7/16-inch nuts and returned them to zero. Then I retightened all 7/16-inch studs to make sure they were secure in the block. Then, I tightened the outboard 3/8-inch nuts to 35 ft-lbs and the 8-mm stud nuts at the inboard locations to 22 ft-lbs and "between lifter" locations. You need to correctly follow the torquing sequence.

It takes some patience to install the four special shouldered washers and torque the 8-mm nuts onto the "extra" four stud locations because of tight space in the lifter valley window. This location is where these studs have been preinstalled to the head deck. When performing this procedure, an effective method is to use a hemostat to place the shouldered washers onto the studs. You need to ensure that the shoulder registers into the hole. You can also use hemostat to locate the nut. Before starting, I rotate the engine block so the head deck is perpendicular to the floor, the lifters parallel to the floor. However, if you rotate too far toward the floor, the lifters can slide out of their bores. Gently begin to rotate the nut onto each stud with one finger.

I then thread all four of these nuts onto their studs. I torque the 8mm nuts to 22 ft-lbs using a 1/4-inch-drive digital torque wrench and a 3/8-inch torque wrench extension that adds 2 inches of length to the torque wrench. Because of the added length of the extension, I set the torque wrench at 19.06 ft-lbs to achieve an actual 22 ft-lbs.

Rockers

Randy Becker from Harland Sharp came to my shop with rockers and pedestals and shims to determine the optimum setup for the Trick Flow L92/LS3 heads. After a bit of test fitting, he concluded that the SL927 rockers (intakes feature an offset) require a rocker pedestal height of .0700 inch to arrive at the proper sweep contact between the rocker roller tip and the valve tips. Randy then machined a set of pedestals to our required height. The machined billet rocker arms feature a 1.7:1 ratio. The supplied 8-mm socket head cap screws were torqued to 22 ft-lbs. Note: Whenever installing LS rocker arm bolts, be aware that all intake rocker arm bolt threads must be coated with a sealing compound such as Teflon thread paste, since the threaded holes in the heads are open to intake ports.

LS3-style cylinder heads require the use of offset intake rockers to accommodate valve size and location. Note the offset intake rocker shown here at left.

Our Harland Sharp full-roller rockers provide superior strength and more precise valve control as opposed to OEM rockers.

Pushrods

During test fitting, I installed light checking springs in the left cylinder head's number-1 intake and exhaust valve positions. Then I rotated the crank to bring the number-1 piston to TDC with the number-1 intake and exhaust cam lobes on their base circles. Next, I bolted on the head with an MLS head gasket and then fitted an adjustable checking pushrod and a pair of our Harland Sharp roller rockers to the casting. Pushrod length came to 7.649 inches at zero lash.

Morel specifies a lifter preload of .045 to .050 inch, so maximum pushrod length would be 7.699 inches. Since pushrods are available in increments of .025 inch, I ordered a set of 5/16-inch chromoly pushrods with .080-inch wall thickness at a length of 7.700 inches from Trend Performance. Trick Flow actually recommends a length of 7.700 inches with its 255-cc LS3 heads, but measuring this length is a smart precaution. I also ordered a few pushrod cleaning brushes, which are handy when cleaning new pushrods to eliminate any manufacturing particles that may be present inside each pushrod's oil passage.

Intake System

For this build, I wanted a straightforward carbureted setup, so I started with a Holley's 300-131 single-plane intake manifold designed for the LS3-style cylinder head. I opted for Holley's 850-cfm Ultra double pumper to handle the fuel management duties. This particular unit has a bright finish with blue anodized base and metering blocks. The 850 Ultra has many exceptional features for high-performance and racing applications. It has an electric choke, mechanical secondary operation, four-corner idle, clear fuel bowl sight glasses, and optimized street/strip calibration. It is 100 percent wet-flow tested in a ready-to-run state.

Designed for LS3/L92 applications, the single-plane intake has 2.50 inches of port height and 1.15 inches of port width. Manifold height is 5.42 inches, which is measured from the block's valley cover to the carb-mounting flange. The intake's ideal operating range is 2,500 to 7,000 rpm. This manifold is also available with machined injector ports for EFI applications and weighs only 11.5 pounds. The manifold intake ports have an O-ring groove around each port to provide a seal against the heads. We used Holley's O-ring seal kit (PN 508-22), and each O-ring seal was carefully installed into the manifold flange grooves. Vaseline or a light coat of lithium grease keeps the seals in position during installation.

The manifold comes with the necessary fasteners: six 6-mm x 1.0 x 50-mm hex-head bolts and four 6-mm x 1.0 x 90-mm hex-head bolts and flat washers. Although these bolts and washers are adequate, I selected ARP polished stainless fasteners. I bought two five-packs of the 6 x 1.0 x 50 bolts, PN 760-1007, and one five-pack of the 6 x 1.0 x 90 bolts, PN 760-1014 (each pack includes washers). I initially torqued the intake mounting bolts coated with ARP moly and then torqued them to 30 in-lbs, 50 in-lbs, then to 106 in-lbs following Holley's recommended tightening pattern.

Note: The intake manifold's bolt holes precisely aligned with the threaded holes in the Trick Flow heads, so no adjustment was required. During early stage test fitting, I checked port alignment of the heads and manifold, and I was duly impressed because the port alignment was close to perfect. Thus, there wasn't any need to machine away material from either the manifold or head ports.

Valve Covers

During the build, I mounted the MSD coils to the valve covers, but I never cared for the appearance of coil brackets. As a result, I chose Holley's new cast-aluminum valve covers. These have mounting bosses, so you can directly mount the coils to the covers and no brackets are needed. Coils are spaced 72 mm apart on the Holley valve cover coil mounts and this requires use of coils applicable to LS2/LS3/LS7 designs. The appropriate MSD coils (PN 82878) are compatible with OEM vehicle applications 2006–2009.

The Holley covers have a stock height and the Harland Sharp roller rockers are somewhat taller than stock and came in contact with the valve covers. Rather than going to taller valve covers that have no coil mounts or require coil brackets, I decided to install spacers. I needed only an additional 3/16 inch or so of clearance, so I decided to use a pair of ICT 3/4-inch-tall billet spacers. While 1/2-inch-tall spacers an option, the 3/4-inch height provides a stouter look and have an O-ring groove for spacer-to-cylinder-head mating. These are compatible with a stock-type gasket or RTV for the spacer-to-valve-cover seal. Two sets of valve cover seals are needed for installation. An OEM-type seal is installed to the O-ring groove side of the spacer and another OEM-type seal is installed to the O-ring groove in the valve cover. I chose a Victor Reinz kit VS50250A (pair of seals and full set of grommets, along with an additional pair of seals PN VS50250).

To fasten the valve covers and the 3/4-inch spacers, I installed stainless-steel socket head cap screws with a length of 3.5 inch, 6-mm stainless flat washers with 18-mm outside diameter, and OEM-type bolt hole grommets. This OD provides a good footprint on the rubber grommets. While Holley instructions recommend a bolt torque of 106 in-lbs, I didn't think this was adequate because the ICT spacers did not include support standoffs. The metal sleeves over OEM valve cover bolts deliver a solid stop between the covers and the heads. Thus, it prevents valve cover roofs from pulling down and distorting. Medium-strength thread locker was applied to the screw threads and the 6-mm screws temporarily tightened to 40 in-lbs. Then I alternated from screw to screw to evenly draw the covers down. The finely machined spacers fit precisely but only included the 6-mm screws and loc washers. The cover spacers should have included length-appropriate sleeves to provide a solid "stop" during tightening, eliminating potential valve cover distortion. Prior to running the engine, I carefully fabricated tube standoffs and I had the Holley valve covers powdercoated in a charcoal wrinkle finish. I also had our front and rear engine covers and oil pan powdercoated as well. I also had the outer walls of the valve cover spacers powdercoated, to visually blend into the valve covers.

Ignition Coils

The Holley valve covers allow you to mount the coils without mounting brackets, which offers an attractive appearance. Coils are secured with 6-mm x 1.0 x 25-mm screws. MSD includes these fasteners with the coils. An upgrade to ARP 12-point polished stainless screws delivered a refined appearance. A bolt spacing kit (PN 82878) for factory LS2/LS7 OEM models was used to fit the MSD coils to the Holley valve covers, which have posts that are spaced 72 mm apart. Tighten the 6-mm bolts to 106 in-lbs.

Trick Flow recommends NGK, PN 4177, spark plugs for this setup. I gapped these at .045 inch.

The only electronic control system needed is for the ignition system because we are using a carburetor for fuel management. An MSD 6LS2 ignition control module with harness kit is wired to the coil harness, water temperature sensor, cam sensor, and crank sensor. The controller features a selection of six preprogrammed ignition curves (simply plug and play), or you can use the included software disc on any PC or laptop computer to tailor your own curve. When removing the fuel injection and installing a carburetor on any LS engine, you simply need the MSD ignition controller kit and you're ready to run.

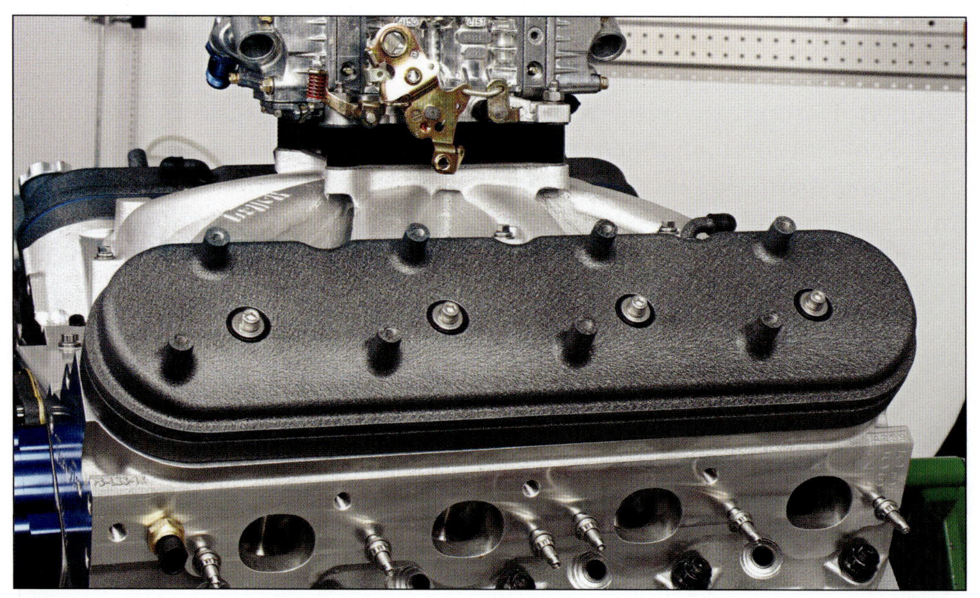

Holley's LS valve covers feature stands that allow direct mounting of MSD coils without the need for unsightly coil brackets.

The LS Next Dyno Session

Gressman Powersports is our favored shop for dyno testing. Scott Gressman owns and runs an updated dyno cell with the latest Superflow dyno. The Moroso oil pan has a 7-quart sump capacity, so we poured in 7 quarts of Brad Penn 30W break-in oil to the sump. Another quart was added for the remote filter and –10 plumbing.

Gressman used the program CD supplied with the MSD 6LS2 controller kit to plot a custom ignition curve. Idle timing was set at 15 degrees and increased to 27 degrees by 2,500 rpm.

Number-80 jets were installed in both primary and secondary metering blocks of the Holley 850-cfm double-pumper carb. When nearly reaching peak power levels, the Superflow air/fuel monitor specified a slightly lean condition, and therefore, Scott installed 90 jets for both. While slightly rich, the engine's power and torque numbers increased. We eventually arrived at number-88 jets for both primary and secondary circuits.

Our brief dyno session measured 665.2 peak hp at 6,000 rpm and 627.3 ft-lbs of maximum torque at 4,900 rpm. Throttle response was extremely snappy and exceptionally quick was the result. Frankly,

I expected the engine to produce peak numbers at marginally higher engine speeds, but if we had been afforded more dyno time, adjusting the carb and ignition timing would likely have produced some further increases. Even so, the engine is definitely an impressive stump puller. Final numbers only provide a vague indicator of output, and of course, on different dynos, numbers can significantly vary. I've actually seen the same engine on different engine dynamometers reading as much as 40- or 50-hp differences (some dynos are simply stingier or more generous).

Block Specs

RPM	HP	Ft-Lbs Torque
3,700	391.7	550.7
3,800	410.8	562.6
3,900	433.2	577.9
4,000	455.3	592.3
4,100	475.2	603.1
4,200	491.8	609.3
4,300	507.9	614.5
4,400	523.4	618.9
4,500	537.5	621.4
4,600	550.4	622.5
4,700	564.3	624.7
4,800	577.6	626.1
4,900	590.8	**627.3**

RPM	HP	Ft-Lbs Torque
5,000	602.1	626.6
5,100	611.1	623.4
5,200	620.0	620.4
5,300	625.6	614.1
5,400	631.3	608.3
5,500	636.1	567.7
5,600	642.1	601.7
5,700	648.0	591.5
5,800	655.2	587.7
5,900	659.9	582.0
6,000	**665.2**	576.8
6,100	662.4	564.9
6,200	660.7	554.4

Dart LS Next Parts List

Dart Next iron block 4.125-inch bore (Iron block with raw 4.000-inch bore also available)	PN 31837211
Original cylinder bore	4.125 inches
Final bore size	4.185 inches
Block deck height	9.240 inches
Approximate lifter bore diameter	.842 inch
Dart cam bearings, coated (specific to LS Next block)	PN 32210101-5
Dart camshaft retainer plate	PN 32226000
Block parts kit (1.5-inch freeze plugs, NPT plugs, head, and bellhousing dowels)	PN 32000016
Crankshaft	SCAT 4.000-inch stroke, PN 4-LS1-1-4000-6125-58
Connecting rods	SCAT 6.125-inch H-Beam, PN 2-350-6125-2100-S
Pistons LS asymmetrical fsr dish/inverted dome 4.185 bore, 1.115 cd, dome volume –14.6	JE PN 324072
Oil pump	Melling PN 10295
Timing set	Cloyes PN 9-3172AZ
Timing chain damper	GM PN 12588670
Crank damper	Fluidampr PN 740111
Mahle Clevite main bearings (Note: Main bearings are specific to LS Next block)	PN MS2321H
Upper rod bearings	Mahle Clevite PN CB-663HNK
Lower rod bearings	Mahle Clevite PN CB-633HXK
Gasket set	Victor PN CS5975
Cylinder heads	Trick Flow PN TFS-3261T004-C01
Cylinder head gaskets (4.200-inch gasket bore; .051inch thick, mls)	Fel-Pro PN 26473L / 26473R
Front cover (powdercoated wrinkle)	Comp Cams PN 5496
Rear cover (powdercoated wrinkle)	GM PN 12639250

Dart LS Next Parts List *Continued*

Valley cover (fabricated aluminum .250 x 6.00 x 20 3/8-inch; with O-ring groove seal)	Birchwood Automotive
Valve covers (powdercoated wrinkle)	Holley PN 241-89
Valve covers spacers (3/4 inch) (powdercoated wrinkle)	ICT Billet PN 641-7
Valve cover gaskets with grommets	Victor PN VS50250A
Extra gaskets for valve cover spacers	PN VS50250
Remote oil filter block fittings (2) (-10 AN Male to –10 AN straight with O-ring)	Summit Racing PN SUM-220164B
Water temperature sensor	Standard PN TX89 (or GM 12608814)
Water temp sensor hole plug for RH head	12-mm x 1.5 x 16-mm bolt (with 12-mm aluminum crush washer and Teflon thread sealant)
Crank position sensor (for 58-tooth)	GM PN 12585546
Cam sensor (2005 and later LS)	GM PN 12591720
Cam sensor wire harness	GM PN 12627501
Camshaft	Comp Cams PN 54-462-11
Lifters (hyd. roller)	Morel/John Callies MP 5294 (.842-inch with V-tie-bars)
Rocker arms	Harland Sharp PN SL927
Pushrods	Trend (7.700 x 5/16 inch with .080 wall)
Crank bolt	ARP PN 234-2503
Rear cover bolts	ARP PN 434-1504
Timing cover bolts	ARP PN 434-1502
Cam bolt kit	ARP PN 134-1003
Carb studs, no spacer	ARP PN 400-2401
Carb studs with 1-inch spacer	ARP PN 400-2402
Oil pump to block 8-mm x 1.25 x 30-mm 5-pack	ARP PN 661-1003
Top cover bolts	ARP PN M10AF20-12
Top cover washers	ARP PN AN25-C516
Interior manifold	ARP PNs 760-1007 and 760-1014
Cylinder head stud kit (head stud kit specific to LS Next block; 23-bolt head application)	ARP PN 234-4341
Crankshaft timing gear key	GM PN 12561513
Intake manifold	Holley PN 300-131
Carburetor	Holley 850-cfm HP PN 82851
Carb 1-inch spacer	Summit PN SUM-G1408
Dipstick assembly	Lokar
Water pump	Meziere PN WP319B
Thermo neck	Meziere PN WN0019B
Oil pan kit (specific to LS Next block. Includes billet spacers and hardware; powdercoated wrinkle)	Moroso PN 20144
Oil pump pickup	Moroso PN 24144
Oil pan gasket	Victor (in Victor gasket set)
Dipstick tube weld-in bung, pan sump (featuring 1/4-inch NPT female thread)	Moroso PN 25970
Remote oil filter fittings	Summit Racing PN SUM-220164 (–10 straight to –10 with O-ring, Qty. 2)
Timing pointer	TCI PN 871005
Belt tensioner	Comp Cams PN 54021
Engine oil	5W40 (specified by Morel lifters)
Ignition controller (for 58-tooth tone wheel)	MSD PN 6LS-2
Coils (with 72-mm bolt spacing LS2/LS3 style)	MSD PN 82878
Spark plug wires (require cutting to length)	MSD PN 32079
Spark plugs (gapped .045 inch)	NGK PN 4177
Oil pump to block bolts	8-mm x 1.25 x 25-mm flange head
Water pump to block	8-mm x 1.25 x 75-mm SHCS
Block paint	Valspar Silver/Sunshades PN 333S439 basecoat/clear

Assembly Procedures

This chapter will provide a number of assembly tips that will aid in the proper final assembly of your LS engine. Rather than outlining every step involved in complete assembly, in this chapter we'll focus on several specific assembly procedures that the first-time LS assembler may or may not already be aware of. Recognizing these specific procedures saves time and aggravation for someone who has not previously assembled the LS platform.

Crankshaft Installation

The main caps on a factory OEM block each feature four vertical primary bolts and one 8-mm "pinch" bolt at the right and left side that engage horizontally from the block into the caps.

If you plan to use GM factory main cap bolts, do not reuse the old bolts, but instead purchase new bolts.

When installing the primary GM main cap bolts, final-tightened using a torque-plus-angle method.

Inboard vertical bolts	15 ft-lbs on the first pass, followed by an additional 80 degrees bolt head rotation
Outboard vertical bolts	15 ft-lbs on the first pass, followed by 53 degrees bolt head rotation
8-mm pinch/cross bolts	18 ft-lbs

The torque sequence of the main cap fasteners is critical to avoid uneven stress that could distort the block. Note that the caps are numbered 1–5 from the front to rear of the block. The main cap tightening sequence is as follows: all vertical inboard primary fasteners are tightened first, followed by the vertical outboard primary fasteners, followed by tightening the 8-mm horizontal pinch, or cross, bolts.

Tighten inboard cap bolts in this order

1. Number-3, driver's side
2. Number-3, passenger's side
3. Number-4, driver's side
4. Number-4, passenger's side
5. Number-2, driver's side
6. Number-2, passenger's side
7. Number-5, driver's side
8. Number-5, passenger's side
9. Number-1, driver's side
10. Number-1, passenger's side

Tighten outboard cap bolts in this order

1. Number-3, driver's side
2. Number-3, passenger's side
3. Number-4, driver's side
4. Number-4, passenger's side
5. Number-2, driver's side
6. Number-2, passenger's side
7. Number-5, driver's side
8. Number-5, passenger's side

Tighten cross bolt cap bolts in this order

1. Number-3, driver's side
2. Number-3, passenger's side
3. Number-4, driver's side
4. Number-4, passenger's side
5. Number-2, driver's side
6. Number-2, passenger's side
7. Number-5, driver's side
8. Number-5, passenger's side
9. Number-1, driver's side
10. Number-1, passenger's side

If aftermarket performance main cap fasteners are selected, follow the fastener maker's tightening specifications. Aftermarket high-performance main cap bolts (or studs and nuts) will commonly require only a torque value, eliminating the need for a torque-plus-angle method. Typical aftermarket main cap fasteners are torqued to 60 ft-lbs at the inboard 11-mm bolts, 50 ft-lbs at the outboard bolts, and 20 ft-lbs for the 8-mm pinch bolts.

Note that the factory main caps feature a slight "wing" protrusion at the ends. It's critical to install the main caps in the proper orientation. The front number-1 main cap's wings must face toward the rear. The rear number-5 main cap's wings must face forward. All remaining caps, numbers 2, 3, and 4, must be installed with the wings facing rearward.

If you're using main studs, you may loosely install the studs prior to installing the caps, as these will serve as guides.

Once all vertical primary main cap fasteners have been fully torqued, install the horizontal 8-mm cross bolts. Once all main cap fasteners are final-tightened, the crank should rotate easily.

Be aware that factory original main caps are directional, and care must be taken to avoid installing them backward. At the end of each main cap, you'll notice a small wing. I've seen first-time assemblers accidentally placing them in the wrong direction, so it's important to point this out. Main cap number-1, the front cap, is installed with the cap wings facing rearward. The rear, number-5 cap, is installed with the wings facing forward. All three remaining caps, numbers 2, 3, and 4, are installed with the wings facing rearward.

GM factory LS blocks tend to react to main cap bolt stress, with potential axial movement of the main bearing bores. After an initial tightening of the primary 11-mm bolts, you may find that the crank does not rotate as easily as you would expect. The bores don't appear to properly align until all main cap fasteners have been fully tightened to specification. I recommend you apply the initial torque value, then gently tap the crankshaft fore and aft using a plastic hammer, rotating the crank between hammer strikes. This, theoretically, provides a better clearance at the center thrust bearing. Only after the final tightening of all 11-mm primary bolts has been done should you tighten the 8-mm side bolts. Once the side bolts have been tightened, the crank should rotate easily. My point is that you shouldn't be surprised if the crank does not rotate easily until *all* main cap fasteners have been fully tightened to spec.

The cross bolts are torqued to 20 to 22 ft-lbs. Tighten each opposing pair at a time: number-3 main cap left side and right side, cap number-4 left and right, etc.

Cylinder Head Installation

Prior to head installation, the lifters must be installed because the installed head gasket and head prevent lifter service.

With the block and head decks clean and dry, install the two head dowel sleeves in the front and rear lower recessed head bolt holes on the block decks. Position the head gasket, noting orientation. High-quality head gaskets should feature a printed "FRONT" to indicate orientation. Double-check to make sure that the cylinder bore holes in the gasket do not encroach into the cylinders. The holes in the gasket must be slightly larger than the cylinder bores.

Carefully position the cylinder head onto the deck, making sure it registers onto the dowel sleeves and that the head mates fully to the gasket and block deck. Examine the entire outer edge to look for any large gaps, which would indicate that the head is not seated properly. Reposition the head by hand. Do not attempt to overcome any uneven gaps by tightening the bolts.

Note: If you are using head studs instead of bolts, loosely finger-install all studs prior to installing the gaskets and heads. Once the head is seated properly, snug each stud to finger-tight or to a maximum of 5–10 ft-lbs. Do not overtighten the studs to the block. Overtightening can result in a slight splay of the studs. When the nuts have been tightened, head clamping is achieved. Depending on the type of fasteners being used, follow GM or aftermarket fastener instructions in terms of bolt lubrication and tightening values. If new GM factory head bolts are being installed, you must use a specific torque-plus-angle tightening method as follows:

11-mm x 2.0 x 155.5-mm bolts	22 ft-lbs plus 76 degrees, plus additional 76 degrees
11-mm x 2.0 x 101-mm bolts	22 ft-lbs, plus 76 degrees, plus additional 34 degrees
8-mm x 1.25 x 46-mm upper pinch bolts	22 ft-lbs

Note that GM factory head bolts are torque-to-yield type and should not be reused. Always install new head bolts.

If aftermarket performance bolts or studs are installed, such as those from ARP or others, follow the maker's instructions, which will involve a torque-only procedure. Be aware that an aftermarket bolt or stud maker may specify a different torque value depending on the thread lubricant. The use of engine oil on bolt threads or on the upper threads of studs will likely require a different torque value than using a specific friction-reducing thread lubricant. If a low-friction lube is applied and the fasteners are tightened according to the value used for oil, overtightening will result. The typical torque value for high-performance head bolts or stud nuts is in the 75–80 ft-lbs specification. Upper 8-mm pinch bolts are tightened to 22 ft-lbs. *Be sure to follow the bolt or stud maker's thread lubrication and torque values for bolts or stud nuts.*

When tightening head bolts, follow the tightening pattern specified by the head manufacturer. Never fully tighten any head bolt to final value in one pass. Following the proper sequence pattern, tighten to an initial value, repeat the same pattern to a higher value, etc., until final value is reached. As an example, when using aftermarket head bolts or studs, I apply an initial torque to all primary fasteners of 20 ft-lbs, followed by another pass at 45 ft-lbs, followed by a pass at 60 ft-lbs, with a final pass at 80 ft-lbs. Only after all primary fasteners are tightened, the upper 8-mm pinch bolts are torqued to 22 ft-lbs.

Pay close attention to your cylinder head gaskets for proper fit before installing the heads. Aftermarket MLS (multi-layer steel) gaskets are usually labeled "FRONT" to indicate orientation, with dedicated left and right gasket part numbers. With the gaskets in position and located onto the block deck dowels, verify that the cylinder holes in the gaskets are slightly larger than the cylinder bores. Gaskets are available for standard and for overbore applications. If the gasket bore edges enter the cylinder bores, sealing will be compromised, the valve may contact the gasket sealing edges, and the exposed edges will create potential hot spots during ignition. Most MLS gaskets carry a special coating that cures under pressure and heat to ensure sealing, so be sure to handle the gaskets with clean hands. The block and head deck surfaces as well as the gaskets must be clean and free of contaminants.

Note that intake rocker bolt holes may be open to intake runners. To avoid potential vacuum leaks, apply a light application of a thread sealer to the intake rocker bolt threads prior to installation.

It's important to note that LS cylinder heads, both GM and aftermarket, feature a 12-mm x 1.5 threaded hole that is open to coolant. LS heads are identical for each bank. This hole

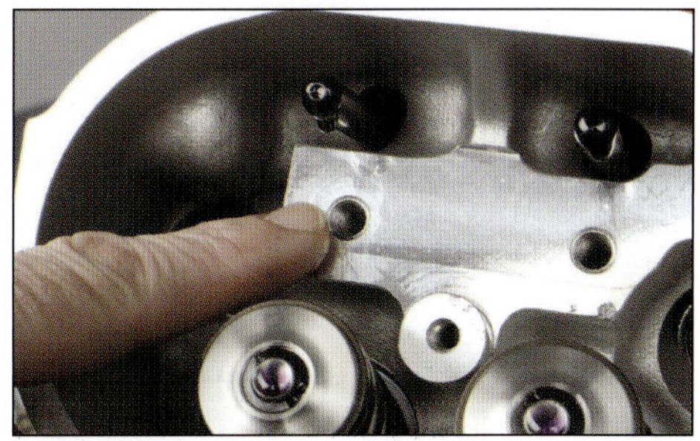

LS cylinder head intake rocker bolt holes may be open to intake runners. This is the case with many aftermarket heads that feature greater runner volume.

is located on the exhaust deck surface toward the end of the head. With heads installed, the left-side head coolant hole will be near the front, just ahead of the number-1 cylinder's exhaust port. The right-side head will feature this hole near the rear, just behind the number-8 cylinder exhaust port. The engine coolant temperature sensor is installed to the left-side head, while the hole on the right head must be blocked off. If you're dealing with original GM heads, the right head should already feature this plug. However, if you're dealing with aftermarket heads, or GM heads that were previously stripped, you'll need a plug to seal the right head's coolant hole. These plugs are not available as separate items through General Motors. However, the fix is easy. Simply obtain a hex-head bolt with thread size 12 mm x 1.5 at a shank length of 15 to 20 mm. Install a copper or aluminum crush washer to the bolt, apply Teflon thread sealant to the bolt threads, and install, tightening to about 25 to 30 ft-lbs.

OEM Lifter Buckets

Factory plastic lifter guides tend to puddle oil at the base of the buckets, slowing down oil return. To eliminate this issue, drill a 5/16-inch hole in the outboard side of each bucket, at each of the four lifter locations, just above the internal floor of each lifter location. The outboard side refers to the side of the lifter guides that faces the exhaust side of the engine. These holes will prevent oil from accumulating inside the buckets and will allow better oil return.

While the factory uses plastic lifter guides to secure individual banks of four lifters in plane to prevent the rollers from rotating, you're certainly not limited to the use of the factory setup. Aftermarket roller lifters are readily available for LS applications that don't require these plastic bucket guides. Lifters are available in pairs, bridged together with a tie bar. The tie bar that connects the pair of lifters maintains the lifter rollers in plane with the cam lobes and features the correct lifter bore-to-bore spacing for LS blocks. This eliminates the need for the OEM plastic guides altogether.

GM plastic lifter guides tend to allow oil to puddle at the base of each lifter pocket. To provide superior oil drainback, drill a 5/16-inch hole in the outboard side of the guide tray, just above the internal floor of the guide.

When installing the factory plastic lifter guides, you must use the factory-type 6-mm screws, which feature a locating shoulder below the screw head to properly locate the lifter set.

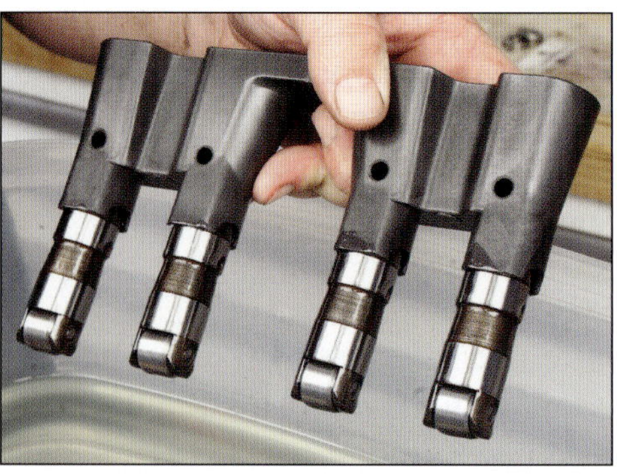

An example of roller lifters snapped into a lifter guide and ready to install. The oil drain holes must face outboard, facing the exhaust side of the engine.

Torque the lifter guide screw to 125 in-lbs. Do not overtighten.

Rear Oil Diverter

Commonly called a "barbell" due to the shape, the rear oil diverter must be installed into the rear of the block, with the O-ring end oriented toward the rear cover. Lightly lube the O-ring and install the diverter by hand until flush with the block surface. Do not forget to install this diverter. If it is left out, you'll lose oil pressure.

Insert the oil diverter until flush with the block rear surface.

Rear Main Seal Installation

A one-piece seal is placed between the crank flange and the rear engine cover on LS crankshafts. Install the rear main seal after the crankshaft main caps have been fully tightened.

The seal installs to the rear engine cover, with the cover and seal installed as a single unit.

The inner seal lips must angle forward for proper positioning. Do not apply lube to either the crank flange or seal lips. Rather, the seal should be installed to the crank flange dry.

New seals may be purchased separately or will be already installed if you purchase a new rear cover. Either way, this will include a large, white nylon seal guide. Insert the seal guide into the seal. As you carefully install the rear cover, the nylon guide will prevent the seal lips from folding rearward. Once the seal is seated onto the crank flange, the guide will fall out. Visually check to verify that the seal lips have not folded rearward. If the lip is not facing forward, remove the cover, reinstall the seal guide, and install the cover again. The seal guide is merely a helper to prevent seal lip fold-back. Once the cover is in place, the guide can be discarded or stored in your tool box for future use.

Finger-install the 8-mm rear cover bolts but do not tighten at this point. Place a straightedge along the oil pan rail and bottom of the rear cover at each side to verify that the rear cover lower faces align with the black pan rails. This minimizes the risk of an oil pan gasket leak. Once oriented properly, tighten the rear cover bolts to 18 ft-lbs in a crisscross pattern.

Front Cover

Front and rear engine covers are secured with 8-mm x 1.25 bolts, torqued to 18 ft-lbs. Installation is very straightforward. While the rear cover should be installed with the main seal dry, apply a smear of oil to the front cover's crank pulley/damper seal prior to installation. An alternative to the factory front cover is Comp Cams' LS cover that offers additional depth to accommodate a double-roller timing chain. This cover also features an eccentric cam sensor bushing that can locate the sensor at the hole's six o'clock position for factory blocks that feature standard cam tunnel height or at twelve o'clock for aftermarket blocks that feature a raised cam tunnel.

Rather than using the factory front cover, Comp Cams makes a very cool aluminum cover that provides clearance for double-roller timing chains and provides a position-adjustable front cam sensor bushing. The bushing features an eccentric hole for the sensor. If you have a factory block, orient the bushing to place the sensor at the six o'clock position. If you have an aftermarket block that features a raised cam tunnel, place the bushing hole at the twelve o'clock position. Cover PN 5496 fits all LS blocks except LS7. PN 5497 fits the LS7 due to that engine's unique oil pump.

Valley Plate

If you're using a multi-port fuel injection system, whether factory or aftermarket, the top engine cover, referred to as a valley plate, is largely hidden from view. However, if you plan to run a carburetor or throttle body EFI atop a carburetor-type intake manifold, the valley plate is visible. While the factory valley plate is certainly viable, you might not prefer its appearance. Aftermarket valley plates are available that are smooth and provide a simplified and cleaner look.

If you don't like what's available, or if you think that an aftermarket plate is too pricey, you can make one using .250-inch-thick aluminum. Width should be 6 inches, and overall length can be as short as 20 inches or as long as 23⅜ inches. Length can vary depending on what you want to achieve at the rear: you can either contour-cut the rear edge to conform to the rear of the block's sealing surface or leave the rear straight and bore a hole to clear the top-rear location of the oil pressure sender. A total of 10 or 11 holes are required for the 8-mm x 1.25 x 20-mm cover mounting screws.

If you wish to add visual interest, the top of the plate can be ball-milled with a series of grooves. The plate can then be polished, plated, anodized, or painted as you prefer. Sealing can be handled either with a factory replacement–style gasket or with a thin bead of RTV. If you want to get fancy, you can have a shallow groove milled to the underside that will accept a length of O-ring seal.

Crank Damper/Pulley Installation

Whether you plan to use a factory crankshaft pulley or an aftermarket crank balancer, *never* use a hammer or another tool to force this onto the crank snout. The damper must be drawn onto the snout using a special installation tool. Apply a light coat of anti-seize lubricant to the snout and to the damper bore. If you're installing a factory pulley, clock orientation is not important. An aftermarket damper may or may not require a locating key. It depends on the combination of aftermarket crank and damper. If a key is featured, hand-position to align the key slot of the damper to the snout key and locate the damper as far as you can by hand. Using a damper installation tool, install a 16-mm x 2.0 thread stub onto the tool and thread the tool into the crank snout. Apply a thin film of anti-seize to the threads first. Draw the damper onto the snout until it bottoms out against the face of the crank gear.

Remove the tool and install the crank snout bolt. If using an aftermarket crank bolt, follow the torque specification. An ARP snout bolt, for instance, is to be tightened to approximately 223 ft-lbs. Since applying torque to the bolt will cause the crank to rotate, the crank may be held steady by installing a pair of temporary crank flange bolts and holding the crank by having a helper place a long bar or long screwdriver between the temporary flange bolts. Note that the ARP crank bolt includes a hardened washer, unlike the GM bolt that features a load-spreading flange as part of the bolt head.

If you're using a GM factory crank bolt, you'll need two of these bolts, one used and one new. With an old (or spare) OEM crank bolt, apply an initial tightening value of 240 ft-lbs. Next, remove this bolt and install a new GM crank bolt. Tighten this new bolt to 37 ft-lbs, followed by additional bolt rotation of 140 degrees. As noted earlier, if using an aftermarket crank bolt, such as those offered by ARP, simply torque to 235 ft-lbs. The ARP crank bolt is also reusable, unlike the GM bolt, which is designed for onetime use only. The GM bolt features a preapplied thread lock compound. The ARP bolt is to be installed using ARP's high-pressure lubricant.

Timing Chain Damper

Depending on the year and version of your LS block, it may or may not have originally included a timing chain damper/rub block. This is a high-impact, lubricity-impregnated composite plastic block that is located between the cam and crank timing gears. It serves to dampen timing chain flutter, and it's always a good idea to have one. Some blocks feature a pair of tapped holes that allow a damper block to directly mount to the block. If your engine block does

The timing chain damper aftermarket bracket and damper mount prior to installing the timing chain. The plastic damper provides a damping for timing chain flutter. If your block is already drilled and tapped for a damper, simply bolt on the plastic damper; otherwise, install the adapter bracket and damper.

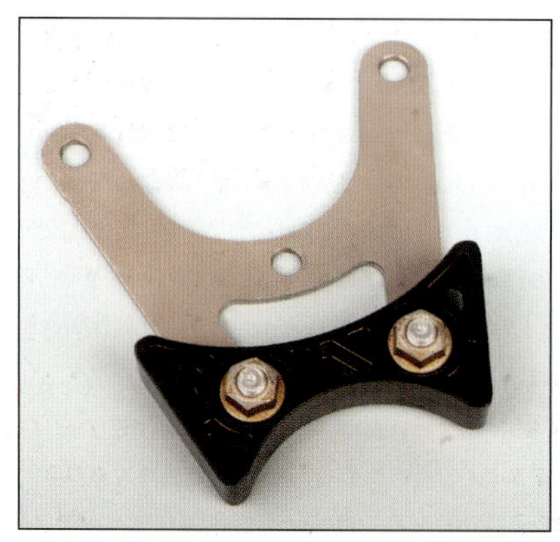

If your LS block is not predrilled and tapped for a timing chain damper block, aftermarket adapter brackets and blocks are available that allow mounting a damper without block modifications, as the bracket shares three bolts at the camshaft retaining plate. Brackets are available from Trick Flow as PN TFS-30675600, and damper blocks are available from GM or as Cloyes PN 9-5417. Melling also offers a bracket and damper kit as PN BD417-DBRKT.

not feature these two holes, don't fret. Aftermarket adapter brackets are readily available that don't require drilling and tapping your block, as the bracket simply shares three of the camshaft retainer plate screws, with no modifications needed. The bracket accepts the plastic damping block. Brackets and rub blocks are available separately or as a kit. The damper/rub block available from all sources is the same as the GM part. Some are black and some are tan, but they're all the same. Sources include GM, Melling, Trick Flow/Summit Racing, Dorman, and others. To my knowledge, the only sources for the adapter bracket are Trick Flow/Summit Racing and Melling. The guys at Trick Flow actually designed it.

Connecting Rod Bolts

The factory specification calls for rod bolts to be tightened to an initial value of 15 ft-lbs, followed by additional angle rotation of 65 degrees or 75 degrees, depending on first or second GM rod bolt design.

Many builders consider the factory rod bolts to be somewhat inadequate, preferring to use high-performance

An excellent alternative to GM rod bolts are high-performance aftermarket rod bolts, available in both metric and inch formats for factory or aftermarket rods. A built-in shoulder centers the bolt, and the strength of these bolts from suppliers such as ARP are superior. Of course, aftermarket performance connecting rods will already include these high-performance bolts.

If using factory connecting rods and new factory rod bolts, follow GM's torque-plus-angle tightening specifications. If using aftermarket rods and bolts, you can follow the maker's torque specification only, or you can monitor rod bolt stretch while creeping up to the torque value. Monitoring bolt stretch allows you to verify both that the bolt has stretched to its desired clamping value and that it has not stretched into its yield limit.

aftermarket rod bolts. Depending on the rods used (factory or aftermarket), follow the rod or rod bolt maker's torque specification. For example, if using Lunati steel forged rods equipped with 7/16-inch rod bolts, their specification is 80 ft-lbs, not to exceed .005- to .0054-inch bolt stretch. I prefer to always monitor rod bolt stretch instead of relying only on a torque value. As an example, I would use a rod bolt stretch gauge to obtain an index of the relaxed bolt length, followed by zeroing the gauge. I then apply an initial torque of 75 ft-lbs, and then connect the stretch gauge to the installed bolt to see how much it has stretched. I then apply additional tightening and recheck to obtain a bolt stretch of .0047 to .005 inch. Using the bolt stretch–checking method eliminates any potential variable in terms of your torque wrench calibration.

Changing Camshafts

If the engine is already assembled but you plan to merely change cams, here is a procedure that will minimize time and labor. *Note that the following procedure only applies to engines that feature factory plastic roller lifter guide buckets.*

1 With the engine cold and coolant drained, remove the water pump and crank damper/pulley.

2 Disconnect the camshaft position sensor connector.

3 Remove the front engine cover.

4 Remove both valve covers.

5 Remove all rocker arms and pushrods, and keep them in order.

6 By hand, rotate the crankshaft two full turns. This will allow all the lifters to be pushed up, away from the cam lobes and snugged into the plastic lifter buckets. The lifters should remain lodged up inside the lifter buckets, clearing a path for cam removal.

However, instead of relying on this, we can install a temporary fail-safe measure to guarantee that the lifters won't accidentally drop.

7 Carefully remove the timing chain and cam gear. At the front of the block, you will notice a lifter galley oil hole above and at the side of the cam.

8 Using a pair of clean steel or aluminum rods, 5/16 inch in diameter and about 22 inches long, carefully insert these rods into the oil galley holes until they gently stop at the rear of the block. In case one or more lifters accidentally drops out of the guide buckets, these rods will serve to keep the lifters out of the path of the cam lobes.

9 Remove the camshaft retaining plate.

10 Carefully remove the old cam.

11 Clean, lube, and carefully install the new cam, being careful to avoid scratching the cam bearings. Once the new cam is fully inserted, with the cam retaining plate installed, remove the temporary rods.

12 Install the cam gear and timing chain, while properly indexing the cam to the crank. Insert a clean pushrod into a lifter's pushrod cup and push downward, popping the lifter out of its high position and onto the cam lobe. Repeat this for all lifters.

13 Reassemble. Keep in mind that depending on the new cam lift, you may need to obtain a different pushrod length, a common issue when you change cams in any pushrod engine.

GM camshaft retainer plates feature an imprinted elastomer seal for the front lifter galley oil holes. These seals are intended for onetime use only. Always install a new cam retainer plate.

Note: There is no need to remove the cylinder heads, lifters, or intake manifold to change cams, unless of course you're making other changes such as upgrading to different heads, manifold, lifters, etc.

As mentioned earlier, the above procedure applies only to the use of GM factory plastic lifter guides. If your build features aftermarket lifters that are tie-bar connected in pairs, the above procedure does not apply. Changing camshafts would require removal of the cylinder heads, head gaskets, and lifters.

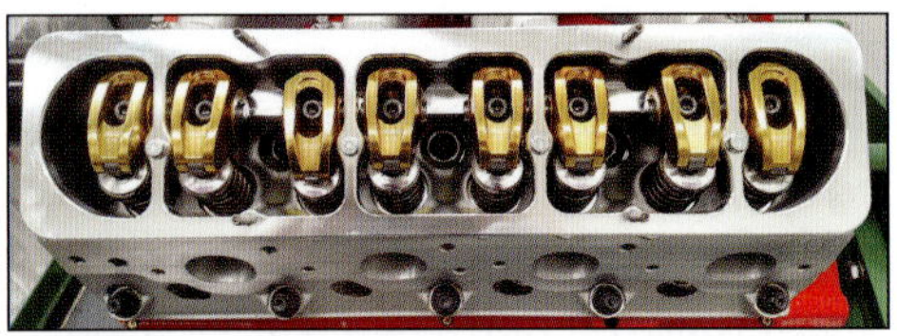

Here's an example of an installed Comp Cams valve cover adapter. Even with a set of fat aluminum roller rockers, I had no clearance issues. A common four-bolt perimeter small-block Chevy valve cover gasket provides the seal between the adapter and cover.

For those who dislike the appearance of coil packs mounted atop valve covers, being able to use small-block Chevy covers opens the door to design ideas, since you have a broad choice among various valve cover styles and heights. I had these tall aluminum SBC covers powdercoated in a wrinkle finish and computer engraved.

Granted, an old-school appearance isn't for everyone, but it's easy to "disguise" an LS if you so desire.

If a fully assembled engine requires a camshaft change, there is no need to remove the intake, heads, or lifters. With the rocker arms and pushrods removed, rotate the crankshaft two full revolutions. This will cause the camshaft lobes to push the lifters up into a "snapped" position in the plastic guides, keeping the lifters out of the way of the cam lobes. Once the crank pulley, water pump, front cover, cam gear, timing chain, and cam retainer plate have been removed, it's a good idea to insert a pair of steel rods into the lifter oil galleys. They will prevent any of the lifters from accidentally dropping out of their guides. Once you remove the old cam, lube and insert the new cam, and remove the rods. Use a pushrod to pop the lifters out of their locked positions, back onto the cam lobes. Use of the metal rods is not mandatory, but serves as a fail-safe to eliminate the potential for lifters to drop while servicing the cam.

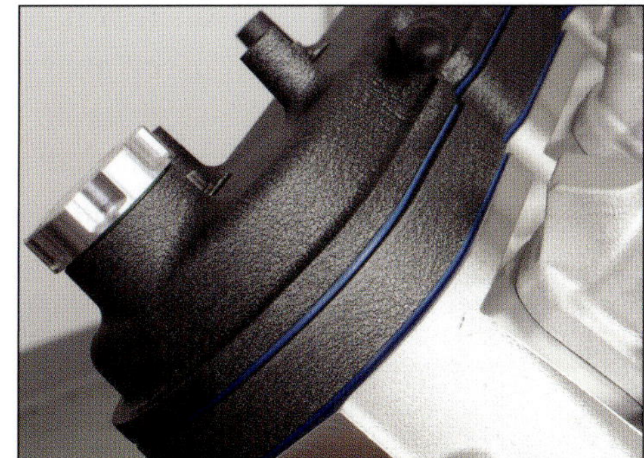

If you're using factory valve covers or aftermarket ones that mimic the originals and provide bosses for coil mounting, you may run into a clearance issue if you're running larger rocker arms. LS valve cover spacers are available in various thicknesses. They simply sandwich between the head and cover, only requiring an additional gasket seal and longer mounting screws.

While GM factory ignition coils are certainly viable, upgrading to high-performance coils, such as those offered by MSD, provides superior and consistent power and improve appearance as well.

Depending on the specific aftermarket intake manifold, sealing is handled with OEM-style gaskets or individual-port O-ring seals.

Valve Cover Swap

Factory LS valve covers are either ugly or pretty, depending on personal preference. If you prefer an old-school appearance, it's easy to swap out the factory LS-style covers in favor of small-block Chevy covers. Available through Comp Cams, Inglese makes a set of very high-quality billet-aluminum adapters that allow you to easily install small-block Chevy valve covers to any LS head.

The adapters bolt to the heads using the stock valve cover threaded holes and seal to the head using standard LS valve cover seals, with the seal between the adapters and Chevy small-block valve covers handled by four-bolt perimeter gaskets. The design of the adapters easily clears factory rockers and even clears a set of large forged-aluminum aftermarket rockers. The thickness of the adapters, added to the height of the

Aftermarket LS valve covers are readily available in various finishes, including bare as-cast, polished, or powdercoated in various colors.

If overkill is your taste, even tunnel ram dual-carb intake manifolds are available for LS applications. This 408-ci build features a Holley intake and dual Holley 650-cfm carbs. This build produced 670 hp at 6,500 rpm on the dyno.

small-block covers, provides more than adequate rocker arm clearances for aftermarket rockers. Of course, you'll need to relocate the coil packs, but the aftermarket offers several mounting systems to locate coils to various underhood locations.

Carb Setup on LS

For those who prefer to run a carburetor on their LS engine, the swap is extremely easy. Select and install a carb-style intake manifold, readily available from makers such as Holley and Edelbrock. Mount the carb of your choice in a cfm size appropriate for your displacement and horsepower level.

You don't need computer control for fuel delivery, so you only need to electronically control ignition. Obtain an LS ignition controller from MSD or another brand, install the eight ignition coils in the location of your choosing, and connect the very simple ignition controller harness. You're done.

Tips: Be sure to select an ignition controller that's compatible with your crankshaft's 24- or 58-tooth reluctor wheel. If installing the engine in a vehicle that was originally equipped with an LS engine, you'll also need to eliminate the high-pressure fuel pump in favor of a low-pressure electric fuel pump, regulated to about 5 to 6 psi. It's that simple. Intake manifolds for LS applications are available in a range of versions, including single 4-barrel dual plane, single 4-barrel single plane, dual carb dual plane, and dual carb single plane. You will need an intake manifold, carburetor, ignition controller with harness, and coil harness when switching from fuel injection.

Custom Cylinder Head Steam Plumbing

Anyone who modifies an LS-format engine in a non-OEM manner faces a challenge with regard to the "steam" plumbing at the cylinder heads. LS heads, whether factory or aftermarket, feature small coolant ports (one at each end of each head). These ports are intended to allow air to bleed from the cooling system from the engine to the radiator. While these ports can be plugged using aftermarket or OEM-available steam plugs, the ports should be plumbed to allow air and over-hot coolant to circulate from the heads to the radiator. For instance, plugging off the rear ports poses a risk of creating hot spots that can affect cylinders number-7 and -8. While the factory uses a tubular system to accomplish this, when you're customizing an LS engine (converting to carburetion or replacing the factory intake and fuel injection with an aftermarket system), the factory plumbing system either can't be used or even if it can be used, looks plain ugly.

Many builders choose to simply plug the steam ports to simplify matters, but as noted earlier, this is a bad idea. There's a simple and effective way to accomplish the coolant plumbing while attaining more of a "performance" appearance.

As an example, I recently built a custom engine that featured an aftermarket block along with a pair of Trick Flow LS heads. To properly evacuate air from the LS-style cylinder heads, I installed a –4 AN coolant transfer plumbing system to the cylinder heads, tying the front and rear cylinder head steam holes together to avoid potential overheating at the rear cylinders. In this build, the –4 AN plumbing system terminates at a Moroso water neck that we installed to the front of the engine. This allows the coolant that runs from the heads to bleed off and circulate into the radiator. Depending on the specific configuration of your build, the plumbing can terminate into a thermostat housing, water neck, or to the radiator itself (by means of a threaded bung that can be welded or brazed to the upper radiator neck).

Note: While this example of steam plumbing was installed on an aftermarket block that features a distributor, the plumbing layout is the same on any LS cylinder head application. There's no reason to be intimidated by creating this custom steam plumbing. It's very easy. It involves a selection of –4 AN fittings and hose ends and a length of –4 hose. If you've never assembled AN hose ends before, it's no big deal.

Items Required

The setup I created involves the following components. In this example, I chose Fragola plumbing items, but feel free to use your brand

4	Trick Flow TFS-30600611 steam plug fittings (featuring –4 male nipples)
4	Fragola 109004-BL 90-degree hose ends, –4
5	Fragola 220104-BL straight hose ends, –4
36.25 inch	Fragola 840604 black –4 hose, 6 feet
(The 6-foot length allows extra length if you make a mistake when cutting)	
2	Fragola 498302-BL T-fittings, male/male/female, –4
1	Fragola 482404-BL T-fitting, male, –4
1	Fragola 496321-BL 90-degree female/female coupler, –4
1	Fragola 481605-BL straight fitting, –4 male to 1/4-inch NPT
2	Fragola 498202-BL T-fittings, male/male/female, –4

Creating a custom steam plumbing system for LS cylinder heads is easy using a selection of –4 AN hardware and hose.

Aftermarket steam hole plugs that feature a –4 AN male fitting directly bolt on to the heads.

The fitting block features a short shoulder and O-ring that seals into the cylinder head's steam port.

of choice (Earls, Aeroquip, Russell, XRP, Goodridge, etc.).

Component Locations

In this example, the steam-plumbing layout was performed as follows. (Note that we refer to the right side as the passenger's side, and the left side as the driver's side.)

A Trick Flow TFS-30600611 fitting was installed at each end of each cylinder head.

Each fitting features a –4 AN male fitting. A 6-mm x 1.0 socket head cap screw secures the fitting to the head. A short shoulder fitting with an O-ring inserts into the coolant port hole in the head's threaded hole. Be sure to lightly lube the O-ring and make sure that the base of the fitting block mates flush with the machined surface of the head when installed. Lightly wiggle the fitting while pushing down to fully seat the O-ring shoulder. Tighten the 6-mm screw to about 90 to 100 in-lbs. Verify to make sure that the fitting block is

All LS heads feature a steam port at each end (front and rear).

Overhead view. In our example, note a hose running from front to rear on the right-side head, and right to left heads at the rear, left to right at the front, and a hose running from front left to our water neck.

Hose Assembly Tips

Cutting stainless braided or stainless braid core AN hose with a woven outer sleeve *may* be done with a hacksaw or a cutoff abrasive wheel. To prevent fraying at the cut line, first tightly wrap the hose at the cut line with tape and cut through the tape area. We need to keep the exposed cut free of wire fraying, and the cut must be square. *A much better method* involves the use of a dedicated cutter that is designed to handle braided AN hose. An example is Summit Racing's PN SUM-900040 braided hose cutter, which looks somewhat similar to a long-handled bolt/padlock cutter. This type of snipper will produce a clean cut with a single squeeze and is surprisingly easy to use.

1 Once the hose has been cut to length it must be inserted into the hose end. For the first-time installer, this can prove challenging and sometimes frustrating. In some cases you may be able to wiggle and push the hose into the hose end's collar, but to make things easier, secure the hose end collar onto a bench vise with a pair of aluminum jaws that feature V-cuts designed to secure the hex of a hose end collar.

To simplify the task, I strongly recommend using a hose assembly specialty tool made by Koul Tools available in all AN sizes. (In this case, we're using the –4 tool, which is in Koul's kit PN 468 that includes –4, –6, and –8 tools).

Separate the two halves of the Koul Tool and place the hose end collar into the cavity, orienting the hose end so that the hose entry end is facing the tapered end of the tool. Snap the tool halves together and place the tool in a bench vise with the tapered entry end facing outward.

Insert the hose into the funneled, tapered end of the tool, engaging the hose end. Using inward pressure and a rotating motion, insert the hose until it stops. Remove the tool from the vise, separate the tool, and inspect the hose end inside the female threaded end. The end of the hose should be located just under the threads.

2 Before attempting to assemble the hose end collar to the hose end, wrap a piece of tape around the hose at the base of the collar. This will provide a reference of hose-to-collar depth engagement when you assemble the hose end.

3 Next, secure the hose and hose end onto a bench vise, with the hex of the hose end secured in a pair of aluminum AN vise jaws and the exposed end of the hose end collar slightly protruding from the vise jaws.

4 Lightly lubricate the hose end threads with a thin oil. Insert the hose end nipple into the collar, engaging the nipple into the entry of the hose. Apply a pushing pressure as you engage the threads. Be very careful to maintain a straight entrance to avoid cocking and cross-threading. Once the threads begin to properly engage, use a 9/16-inch open-end wrench (preferably an aluminum –4 wrench) to fully thread the hose end into position. While threading, keep an eye on the reference tape. If the tape begins to excessively move away from the hose end collar, you're pushing the hose out, in which case you'll need to start over. You can expect the tape to move slightly (perhaps 1/8 inch or so), which is fine, as long as the nipple remains seated inside the hose. Once the hose end has been assembled, remove the reference tape from the hose.

5 Before installing a completed hose assembly, make sure that all threads are clean and free of burrs, and lightly lube all male threads to aid in installation. Once the entire system has been installed on the engine, start the engine to allow coolant flow and check each connection for leaks. If you find a leak, lightly back off the threaded connection and retighten. Do not overtighten. Sealing takes place at the hose-end-to-fitting 37-degree seat. Do not apply any type of thread sealing tape or paste to any tapered seat hose end or fitting. Only apply a thread sealer to any NPT threaded connection (for example, where you install an NPT fitting to a water neck housing). ■

Inserting an AN hose into the hose end hex collar can be a chore for the first-time installer. Using Koul Tools's AN hose end assembly tool makes this a breeze.

fully flush to the flat milled mating surface on the cylinder head. Simply tightening the screw more may not work. The short shoulder at the O-ring must be fully engaged with the bottom of the block flush with the mating surface.

At the right front TFS fitting

1. Connect a male/male/female T-fitting, with the female end attached to the TFS fitting.
2. Connect a 90-degree hose end onto the vertical male of the T-fitting.
3. Connect a straight hose end onto the horizontal male of the T-fitting.

At the right rear TFS fitting

1. Connect a male/male/female T-fitting, with the female end attached to the TFS fitting.
2. Connect a 90-degree hose end onto the vertical male of the T-fitting.
3. Connect a straight hose end onto the horizontal male of the T-fitting.

At the left rear TFS fitting

1. Connect a 90-degree hose end to the TFS fitting's –4 male nipple.

At the left front TFS fitting

1. Connect a 90-degree female/female coupler to the TFS fitting.
2. Connect a male/male/male T-fitting to the female coupler.
3. Connect a straight hose end to the horizontal of the male T-fitting.
4. Connect a 90-degree hose end to the vertical male of the T-fitting. This hose end will allow you to feed a –4 hose from this 90-degree hose end to your water neck or radiator neck.

Our –4 Hose Lengths

Right-front-to-right-rear hose	16.5 inches
Rear-right-to-left hose	8 inches
Front-left-to-right hose	4.5 inches
Left-front-to-water-neck hose	7.25 inches

In this build, we ran the hose from the left front 90-degree hose end to our Moroso water neck. The water neck housing features a 1/4-inch NPT threaded hole, where we installed a 90-degree –4 AN to a 1/4-inch NPT fitting, which accepts a straight –4 hose end.

Hose Lengths

Once you have mocked up all fittings and hose ends, you can measure for individual hose lengths. Regardless of your specific engine application, as long as you have LS heads, the right-front-to-right-rear hose and the front-left-to-right hose lengths can duplicate our example. The only hose length variables may include the rear-left-to-right crossover hose and the final outlet hose that runs from the left front to your point of termination.

In our example, the hose length running from the left front to the Moroso water neck was 7.25 inches long. The length of this final hose will vary depending on your point of connection to a water neck or upper radiator neck (which will likely require brazing a 1/4-inch NPT bung to the radiator neck).

Note: This build featured an aftermarket block that is based on small-block Chevy Gen I architecture designed to accept LS cylinder heads. It features a Chevy-style rear-mounted distributor. Our right-rear-to-left-rear crossover hose was 8 inches long, allowing a slight curve to clear our distributor. If you're running an LS block (no rear distributor), this right-to-left-rear hose can be slightly shorter.

Creating the system requires four separate hose assemblies.

LS Fastener Torque Specifications

Included here are both factory OEM and examples of specific aftermarket specifications. Applying the correct tightening values for threaded engine fasteners is critical for any engine build. Listed here are both factory OEM values, some of which involve a torque-plus-angle tightening method, and OEM and aftermarket values that require only a tightening-by-torque approach. Especially with regard to main cap bolts, cylinder head bolts, and connecting rod bolts, it is imperative to follow the correct tightening procedures depending on the use of OEM bolts or aftermarket high-performance fasteners. Each requires a different tightening method. These OEM bolts require a torque-plus-angle procedure while aftermarket fasteners require only a torque value. In addition, the OEM main cap, cylinder head, and rod bolts are torque-to-yield, so they should not be reused. This also holds true for the crankshaft bolt that secures the crankshaft pulley. The OEM crank bolt is designed as a onetime-use bolt and should never be reused.

OEM Factory Torque Values

Always check with the service manual to verify because some values may change depending on engine model.

Main Cap Bolt Torque Values (OEM)	
Inner main cap bolts	15 ft-lbs (first pass), plus 80 degrees final
Outer main cap stud nuts	15 ft-lbs (first pass), plus 53 degrees final
Main cap side bolts	18 ft-lbs

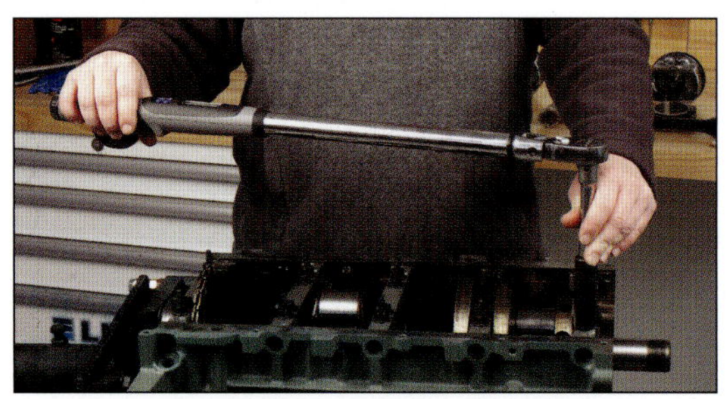

Following the correct bolt tightening sequence, if using factory main cap bolts, first tighten the inboard main cap bolts to 15 ft-lbs. Next, tighten all outboard main cap bolts to 15 ft-lbs. Next, angle-tighten the inboard bolts by an additional 80 degrees, followed by angle-tightening the outboard bolts by an additional 53 degrees. Finally, tighten the 8-mm main cap side bolts to 18 ft-lbs. If using aftermarket main cap bolts, such as those offered by ARP, tighten main cap inboard and outboard bolts to 80 ft-lbs (tighten in three incremental steps, following the correct tightening sequence in each step, finalizing at 80 ft-lbs). No additional angle tightening is required. Finally, tighten the 8-mm side bolts to 22 ft-lbs. If using ARP main cap studs, install the studs finger tight to the block and apply tightening torque to the nuts.

Do not install the main cap 8-mm side bolts until all primary main cap bolts have been fully installed.

Once the 8-mm main cap side bolts have been torqued, the crank should rotate easily. Do not be surprised if the crank does not turn easily until these side bolts have been installed.

It is highly recommended to upgrade all main cap and head bolts to aftermarket fasteners such as those offered by ARP. These aftermarket bolts offer superior strength and can be reused.

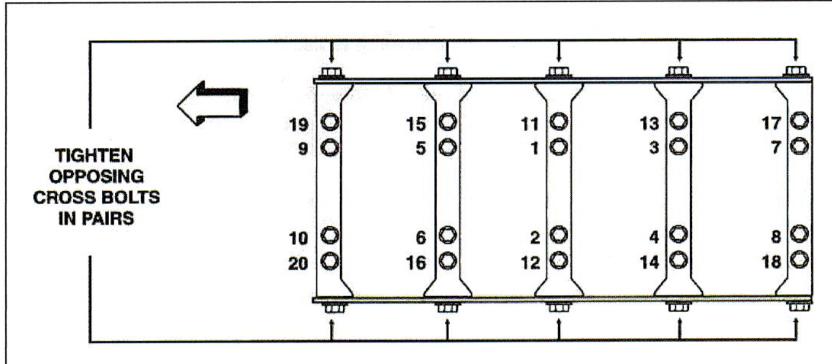

Follow this torquing sequence for LS main cap bolts.

When tightening cylinder head fasteners, always follow the proper tightening sequence to evenly spread the clamping load across the head. Also, tighten in steps instead of applying full torque in one step; for example, when using aftermarket performance head bolts, make the first pass at 20 ft-lbs, followed by a second pass at 45 ft-lbs and a final pass at 75 ft-lbs.

Connecting Rod Bolts (OEM)	
OEM rod bolts (first design)	15 ft-lbs first pass, plus 65 degrees second pass
OEM rod bolts (second design)	15 ft-lbs first pass, plus 75 degrees second pass

Exhaust Manifold	
Exhaust manifold bolts first pass	11 ft-lbs
Exhaust manifold bolts final pass	18 ft-lbs
Exhaust manifold heat shield bolts	80 in-lbs

Intake Manifold	
OEM aluminum	106 in-lbs
Plastic	44 in-lbs, final at 89 in-lbs

Cylinder Head Bolt Tightening (OEM)	
11-mm x 2.0 x 155.5-mm bolts	22 ft-lbs, plus 76 degrees, plus 76 degrees
11-mm x 2.0 x 101.0-mm bolts	22 ft-lbs, plus 76 degrees, plus 34 degrees
8-mm x 1.25 x 46-mm bolts	22 ft-lbs
Note: Original equipment cylinder head bolts are torque-to-yield type and should not be reused. If planning to use OEM cylinder head bolts during a build, always use new bolts.	

Miscellaneous OEM Bolt Torque Specs	
Camshaft retainer plate bolts	18 ft-lbs
Camshaft gear bolts	26 ft-lbs
OEM timing chain damper bolts	18 ft-lbs
Oil pump mounting bolts	18 ft-lbs
Oil pump cover bolts	106 in-lbs
Oil pump relief valve plug	106 in-lbs
Oil pump screen nuts	18 ft-lbs
Oil pump to screen bolt	106 in-lbs
Oil pan bolts 8 mm (pan to block and pan to front cover)	18 ft-lbs
Oil pan bolts 6 mm (pan to rear cover)	106 in-lbs
Oil filter fitting	40 ft-lbs
Oil level indicator tube bolt	18 ft-lbs
Oil level sensor	115 in-lbs
Oil pan baffle bolts	106 in-lbs
Oil pan closeout cover bolts	106 in-lbs
Oil pan cover bolts	106 in-lbs
Oil pan drain plug	18 ft-lbs
Oil pressure sensor	15 ft-lbs
Oil filter	22 ft-lbs
Knock sensors	15 ft-lbs
Front cover bolts	18 ft-lbs
Rear cover bolts	18 ft-lbs
Block valley cover bolts	18 ft-lbs
Water pump bolts	11 ft-lbs first pass, 22 ft-lbs final pass
Water pump cover bolts	11 ft-lbs
Lifter bucket 6-mm bolts	125 in-lbs
Flywheel bolts	15 ft-lbs first pass, 37 ft-lbs second pass, and 74 ft-lbs final
Spark plugs	15 ft-lbs (new: 11 ft-lbs with used plugs)
Throttle body bolts	106 in-lbs
Transmission housing (bellhousing) bolts	37 ft-lbs
Valve lifter guide tray bolts	106 in-lbs
Rocker arm bolts	22 ft-lbs
Rocker arm cover bolts	106 in-lbs
Water inlet housing bolts	11 ft-lbs
Crankshaft balancer bolt initial seating	240 ft-lbs
Crankshaft balancer bolt (using new bolt)	37 ft-lbs plus 140 degrees
Crankshaft oil deflector nuts	18 ft-lbs
Crankshaft position sensor bolt	18 ft-lbs
Camshaft sensor bolt	106 in-lbs
Coolant temperature sensor	15 ft-lbs
Cylinder head coolant plug	15 ft-lbs
Cylinder head core hole plug	15 ft-lbs
Block coolant drain plugs	44 ft-lbs
Coolant air bleed pipe bolts	106 in-lbs
Engine mount heat shield nuts	89 in-lbs
Engine mount through bolts	70 ft-lbs
Engine mount through bolt nuts	59 ft-lbs
Engine mount to engine block bolts	37 ft-lbs
Engine service lift bracket 10 mm bolts	37 ft-lbs
Engine service lift bracket 8 mm bolt	18 ft-lbs
EGR valve bolts first pass	89 in-lbs
EGR valve bolts final pass	22 ft-lbs

Miscellaneous OEM Bolt Torque Specs	CONTINUED
EGR valve-pipe-to-cylinder-head bolts	37 ft-lbs
EGR valve-pipe-to-exhaust-manifold bolts	22 ft-lbs
EGR valve-pipe-to-intake-manifold bolts	89 in-lbs
Fuel injection rail bolts	89 in-lbs
Alternator bracket bolts	37 ft-lbs
Alternator rear-bracket-to-block bolt	18 ft-lbs
Alternator rear-bracket-to-alternator bolt	18 ft-lbs
Ground strap bolt at rear of cylinder head	37 ft-lbs
Ignition-coil-to-bracket bolts	106 in-lbs
Ignition-coil-bracket-to-valve-cover bolts	106 in-lbs
Drive belt idler pulley bolt	37 ft-lbs
Drive belt tensioner bolts	37 ft-lbs
Air injection reaction pipe to exhaust manifold	15 ft-lbs
A/C compressor bolts	37 ft-lbs
A/C compressor bracket bolts	37 ft-lbs
A/C idler pulley bolt	37 ft-lbs
A/C tensioner bolt	18 ft-lbs
Accelerator control cable bracket bolts	89 in-lbs

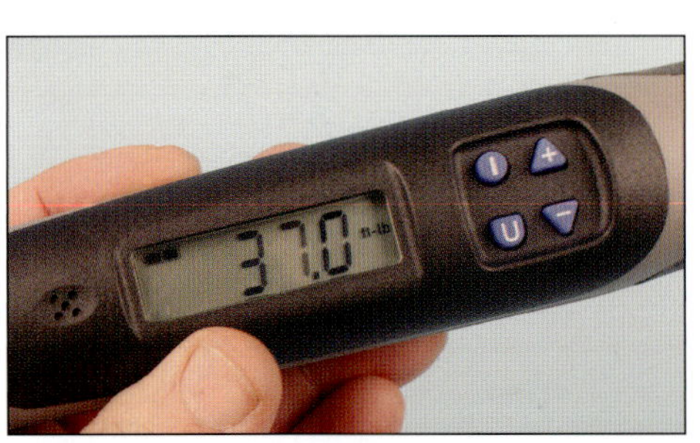

If using factory OEM main cap and/or cylinder head bolts that require a torque-plus-angle tightening method, you'll need to monitor the degrees of bolt head rotation once the bolts have been initially torqued. Available methods include placing a dot on the bolt head and observing how far the bolt head is turned, using an inexpensive degree wheel to your wrench, and using a professional-grade digital torque wrench that provides both torque and degree modes. This Snap-On digital torque wrench, for example, allows you to enter the torque mode and set your initial torque.

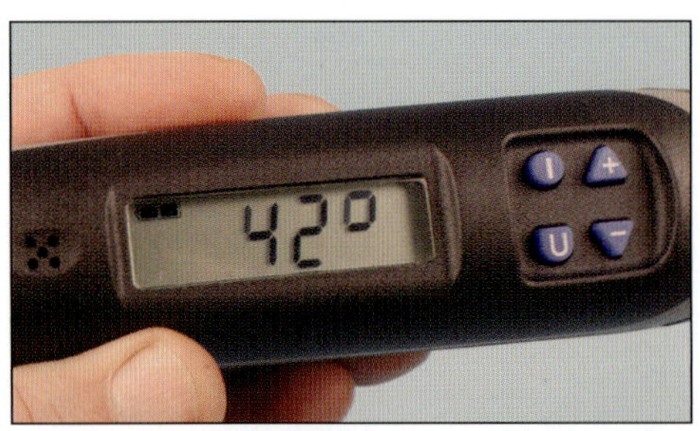

Using the Snap-On digital wrench as our example, once the bolts have been initially tightened by torque, enter the degree mode and set the number of degrees required. Once you reach the desired torque or the desired number of degrees during tightening, the wrench provides an audible beep. One handy feature of the Snap-On digital wrench is that you can tighten by ratcheting even during angle tightening.

OEM Block Fastener Sizes	
OEM LS engines use 100-percent metric fasteners.	
Rear of block (for engine stand mounting): 10 mm x 1.5 x length as needed. (Length as needed according to your engine stand. Our Goodson engine stand requires a length of 120 mm.)	
Cylinder head bolts (OEM)	11 mm x 1.5 x 100 mm
Cylinder head pinch bolts	8 mm x 1.25 x 45 mm
Main cap primary bolts (inner)	10 mm x 2.0 x 100 mm
Main cap bolts with stud tips (outer)	10 mm x 2.0 x 85 mm
Main cap side bolts	8 mm x 1.25 x 25 mm
Timing cover bolts	8 mm 1.25 x 30 mm
Rear cover bolts	8 mm x 1.25 x 25 mm
Lifter tray retaining bolts	6 mm x 1.0 x 20 mm (shouldered)
Camshaft plate screws	8 mm x 1.25 x 25–30 mm
Valley cover plate bolts	8 mm x 1.25 x 30 mm
Timing chain dampener bolts	8 mm x 1.25 x 35 mm
PN 12561663 left-side water plug	30 mm x 1.25 x 10 mm
PN 11588949 straight-thread plugs	16 mm x 1.5 x 10 mm
PN 9427693 front oil expansion plug	16–mm diameter
Connecting rod bolts	9 mm x 1.0 x 43 mm
Crankshaft snout damper bolt	16 mm x 2.0 x 105 mm

Aftermarket Component Torque Values

Performance aftermarket fasteners commonly provide a torque specification only, instead of an OEM specification that may have included a torque-plus-angle value. Common OEM applications for torque-plus-angle include crankshaft balancer bolt, main caps, rod bolts, and head bolts. Always check with the instructions for torque value of the specific component, as aftermarket manufacturer specifications may vary.

Listed on the following pages are examples for specific components. Most remaining fasteners will require the same values as OEM.

Note that many aftermarket performance engine blocks feature inch thread sizes instead of metric sizes for areas such as main caps and cylinder heads.

The torque values are specific to aftermarket fastener use. Torque values for other components such as front and rear engine covers, oil pump, etc., retain the OEM values.

If using ARP bolts, apply ARP fastener lube to the underside of all bolt heads.

Also apply ARP lube to the washers.

Apply ARP assembly lube to the threads. ARP's tightening torque specifications are based on the use of this lubricant. This aids in reducing the friction variable to achieve a more accurate and consistent clamping load.

The OEM crank pulley may be initially seated using an old factory crank bolt, which is then discarded in favor of a new factory bolt. In lieu of using a factory bolt, a 16-mm x 2.0 threaded rod and an adapter may be used to interference-fit the pulley to the crank snout. If using an aftermarket balancer, a balancer installation tool should be used. Once the pulley or balancer has been seated, a new factory crank bolt or an ARP crank bolt may be installed.

Aftermarket Main Cap Studs	
ARP inboard primary main cap nuts	60 ft-lbs
ARP outboard primary main cap nuts	50 ft-lbs
ARP 8-mm main cap side bolts	20 ft-lbs
ARP crankshaft nose bolt (for crank pulley/ balancer)	235 ft-lbs

Note: When using aftermarket main studs, always follow the stud maker's tightening specifications for its studs and nuts. Note that some aftermarket blocks feature four-bolt main caps with no provision for the 8-mm side bolts. Also, torque specifications may be listed for oil or a specific low-friction lube on threads. The values listed here are for use with ARP lube. Values may be slightly higher if oil is applied. Pay attention to the fastener maker's specifications.

A vast improvement over the factory crank bolt is ARP's crank bolt. While the factory bolt is intended as a onetime-use bolt, the ARP bolt may be reused. Tightening the factory crank bolt requires an initial seating torque of 240 ft-lbs. The bolt is then loosened, then torqued to 37 ft-lbs, followed by an additional angle tightening of 140 degrees.

In contrast, the ARP bolt is simply torqued to 235 ft-lbs, using ARP's Ultra assembly lube. Note that an old factory crank bolt may be used to initially seat the crank pulley. The bolt should then be discarded and replaced with a new factory crank bolt, torqued to 37 ft-lbs and then turned an additional 140 degrees as mentioned previously. Yes, the ARP crank bolt is more expensive than the factory bolt, but because of the advantage of installation procedure and superior durability, the upgrade is highly recommended.

While you may use an old factory crank bolt to seat the crank pulley, for final installation, a new ARP crank bolt is highly recommended. Unlike a factory crank bolt, the ARP bolt requires only a torque value and may be reused.

Aftermarket Rod Bolts

Rod bolt sizes and torque values vary among manufacturers. Always follow the rod and/or rod bolt maker's specifications.

Example: Lunati forged rods with 7/16-inch connecting rod bolts: .005 to .0054–inch stretch or 80–85 ft-lbs.

Some LS factory rod bolts feature a bushing sleeve under the bolt head designed to properly center the bolt in the rod cap and rod. The interference fit of these bushings is not reliable. ARP offers upgraded rod bolts that feature a larger diameter shoulder to eliminate the need for the sleeve. Also, while OEM rod bolts require torque-plus-angle tightening, the ARP bolts require only a torque application and can be reused. Upgrading to ARP rod bolts is highly recommended for any LS build. Shown here is a factory rod bolt (left) next to an ARP rod bolt.

Note: When using aftermarket connecting rod bolts, be sure to follow the bolt manufacturer's specific tightening specifications, which might include torque, torque-plus-angle, or bolt stretch specs.

If you're using aftermarket performance connecting rods, you cannot use factory-style rod bolts. The rods will include high-performance rod bolts that are matched to the rods in terms of thread diameter, thread pitch, length, and tensile strength. Aftermarket performance rods for LS applications typically feature 7/16-inch-diameter rod bolts. Since torque specifications vary depending on thread size and bolt design, it's critical to follow the torque specifications provided by the rod maker.

When removing connecting rod bolts from a loose rod, or when tightening rod bolts to a loose rod for bearing clearance checking, the rod big end must be secured. Instead of using a common bench vise, the correct method is to secure the rod in a dedicated rod vise that protects the rod from gouging or distorting.

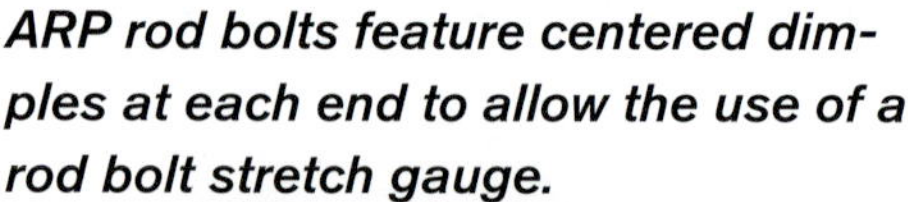

ARP rod bolts feature centered dimples at each end to allow the use of a rod bolt stretch gauge.

If high-performance connecting rods are being installed, the specifications card included with the rod set will provide both a torque value and a maximum bolt stretch value. If tightening by monitoring rod bolt stretch, you'll need a rod bolt stretch gauge, which is available from a variety of sources, including ARP, Goodson Supply, Gearhead, and others. The aftermarket rod bolts feature a dimple at both the head and the shank tip. Install the bolt to the gauge and adjust the dial to zero. This provides a reference of the bolt's relaxed free length. Once the bolt has been installed and torqued, the gauge is positioned onto the bolt to determine how far the bolt has stretched.

High-performance aftermarket connecting rod bolts, such as those made by ARP, can be installed by torquing to the maker's specifications based on the bolt diameter and grade or by tightening following a bolt stretch monitoring method. The bolt maker will specify the maximum allowable stretch for a given bolt. The bolt needs to stretch by a certain amount to provide a "spring-like" clamping load but must not be stretched beyond its elastic limit.

When tightening rod bolts, or for any bolt, never jerk the torque wrench. Always apply an even, steady, and slow pull for improved accuracy and bolt-to-bolt consistency.

Aftermarket Cylinder Head Stud Nuts	
ARP cylinder head stud 7/16-inch nuts	80 ft-lbs with ARP moly
(The 7/16-inch nuts are tightened in three steps.)	
ARP 8-mm upper studs	22 ft-lbs with ARP moly

If you opt to use studs for main caps or cylinder heads, do not apply a great deal of torque when installing the studs into the block. Install studs to the block finger tight, or at the most with about 5 ft-lbs of torque. The main caps and/or cylinder heads will be secured by tightening the nuts to the studs. Overtightening the studs into the block can create a slight splayed condition, affecting stud alignment and potentially reducing the even footprint of the nut and washer.

If the rockers of choice are nonadjustable, whether OEM or aftermarket, tighten rocker arm bolts to 22 ft-lbs.

Rocker Arms

If installing OEM or aftermarket rocker arms that are nonadjustable, the rocker arm bolts that secure the arms to the cylinder heads are torqued to 22 ft-lb.

Aftermarket Water Pump

Torque to 18 ft-lbs.

Thread Condition

As part of your engine block preparation, inspect all threaded bolt holes for cleanliness and thread condition. Critical threaded holes especially, such as those for main cap bolts and cylinder head bolts, must be in good condition to achieve correct and consistent clamping loads. While you may be tempted to use a common "cutting" tap to clean these threads, avoid this at all cost. A cutting tap may allow easy insertion of bolts, but this type of tap will remove thread material, which can weaken the fastener connection. Instead, use a dedicated forming tap, also referred to as a follower, or cleaner, tap. This style of tap is designed to reshape or reform the threads without removing material. Forming taps for LS applications are available from sources such as ARP.

During block preparation and prior to assembly, the block's threaded holes for the main caps and the cylinder heads should be verified for both cleanliness and for thread condition. Any contaminants and/or thread burrs will adversely affect bolt installation and fastener torque, and critical clamping loads as a result. Do not use a conventional cutting tap to clean these threaded holes. A cutting tap will remove material and potentially weaken the threads. Instead, use a dedicated forming tap, such as those available from ARP specifically for the LS engine block. A forming tap will reform/reshape the threads without removing critical material. Shown here is an 11-mm forming tap for cylinder head bolt holes.

The LS engine platform is well supported by the performance aftermarket. All components, including blocks, heads, and all assembly components, are available as upgrades for high-performance and racing applications to boost both performance and durability. The following list is not engine generic. The manufacturers listed here all offer components that apply to LS engine applications.

Engine Blocks

BMP
420 Hamilton Rd., Unit A
Deland, FL 32724
386-279-7131
billmitchellproducts.com

Chevrolet Performance
6200 Grand Pointe Dr.
Grand Blanc, MI 48439
800-450-4150
chevroletperformance.com

Dart Machinery
353 Oliver St.
Troy, MI 48084
248-362-1188
dartheads.com

ERL Performance
2560 Charlestown Rd.
New Albany, IN 47150
877-815-3434
erlperformance.com

RHS
3416 Democrat Rd.
Memphis, TN 38118
877-776-4323
racingheadservice.com

World Products
7301 Global Dr.
Louisville, KY 40258
877-630-6651
pbm-erson.com

Crankshafts

Bryant Racing
1600 E. Winston Rd.
Anaheim, CA 92805
714-535-2695
bryantracing.com

Callies Performance Products
PO Box 926
Fostoria, OH 44830
419-435-2711
callies.com

Dart Machinery
353 Oliver St.
Troy, MI 48084
248-362-1188
dartheads.com

Eagle Specialty Products
8530 Aaron Ln.
Southaven, MS 38671
662-796-7373
eaglerod.com

Kings Crankshaft
PO Box 662
Denver, NC 28037
704-483-1005
kingscrankshaft.com

Lunati
8649 Hacks Cross Rd.
Olive Branch, MS 38654
662-892-1500
lunatipower.com

Manley Performance Products
1960 Swarthmore Ave.
Lakewood, NJ 08701
800-526-1362
manleyperformance.com

Scat Enterprises
1400 Kingsdale Ave.
Redondo Beach, CA 90278-3983
310-370-5501
scatcrankshafts.com

Winberg Crankshafts
333 W. 48th Ave.
Denver, CO 80216
303-783-2234
winbergcrankshafts.com

Connecting Rods

Brian Crower
161 S. Marshall Ave.
El Cajon, CA 92020
619-749-9018
runbc.com

Callies Performance Products
PO Box 926
Fostoria, OH 44830
419-435-2711
callies.com

Eagle Specialty Products
8530 Aaron Ln.
Southaven, MS 38671
662-796-7373
eaglerod.com

GRP Connecting Rods
333 W. 48th Ave.
Denver, CO 80216
303-935-7565
grpconrods.com

K1 Technologies
7201 Industrial Park Blvd.
Mentor, OH 44060
800-321-1364
k1technologies.com

Lunati
8649 Hacks Cross Rd.
Olive Branch, MS 38654
662-892-1500
lunatipower.com

Manley Performance Products
1960 Swarthmore Ave.
Lakewood, NJ 08701
800-526-1362
manleyperformance.com

MGP Connecting Rods
1560 Tuskegee Pl.
Colorado Springs, CO 80915
719-219-3107
mgpconnectingrods.com

Oliver Racing Parts
5339 M-66 North
Charlevoix, MI 49720
231-237-4515
oliverracingparts.com

Venolia Pistons
2160 Cherry Industrial Cir.
Long Beach, CA 90805
562-531-8463
venolia.com

Pistons

Arias Pistons
13420 S. Normandie Ave.
Gardena, CA 90249-2212
310-532-9737
ariaspistons.com

CP-Carrillo
1902 McGaw Ave.
Irvine, CA 92614
949-567-9000
cp-carrillo.com

Diamond Racing Products
23003 Diamond Dr.
Clinton Twp., MI 48035
877-552-2112
diamondracing.net

Hypermax Engineering
255 E. Rt. 72
Gilberts, IL 60136-9627
847-428-5655
gohypermax.com

Icon Pistons
1040 Corbett St.
Carson City, NV 89706
800-648-7970
uempistons.com

JE Pistons
10800 Valley View St.
Cypress, CA 90630
714-898-9764
jepistons.com

Katech
24324 Sorrentino Ct.
Clinton Twp., MI 48035
586-791-4120
katechengines.com

Mahle Motorsports
270 Rutledge Rd.
Fletcher, NC 28732
888-255-1942
mahlemotorsports.com

Manley Performance Products
1960 Swarthmore Ave.
Lakewood, NJ 08701
800-526-1362
manleyperformance.com

Probe Industries
2555 W. 237th St.
Torrance, CA 90505
310-784-2977
probeindustries.com

Ross Racing Pistons
625 S. Douglas St.
El Segundo, CA 90245
310-509-0100
rosspistons.com

Speed-Pro/Federal Mogul
26555 Northwestern Hwy.
Southfield, MI 48033
248-354-7700
speedpro-pistons.com

Venolia Pistons
2160 Cherry Industrial Cir.
Long Beach, CA 90805
562-531-8463
venolia.com

Wiseco Performance Products
7201 Industrial Park Blvd.
Mentor, OH 44060
440-951-6600
wiseco.com

Camshafts

Bullet Racing Cams
8785 Old Craft Rd.
Olive Branch, MS 38654
662-893-5670
bulletcams.com

Callies Performance Products
PO Box 926
Fostoria, OH 44830
419-435-2711
callies.com

Competition Cams
3406 Democrat Rd.
Memphis, TN 38118
901-795-2400
compcams.com

Crane Cams
1830 Holsonback Dr.
Daytona Beach, FL 32117
866-388-5120
cranecams.com

Crower Cams & Equipment
6180 Business Center Ct.
San Diego, CA 92154-5604
619-661-6477
crower.com

Edelbrock
2700 California St.
Torrance, CA 90503
310-781-2222
edelbrock.com

Erson Camshafts
7301 Global Dr.
Louisville, KY 40258
800-641-7920
pbm-erson.com

Isky Racing Cams
16020 S. Broadway
Gardena, CA 90248
iskycams.com

Lunati
8649 Hacks Cross Rd.
Olive Branch, MS 38654
662-892-1500
lunatipower.com

Ultradyne Racing Cams
(see Bullet Cams)

Cylinder Heads

Air Flow Research
28611 W. Industry Dr.
Valencia, CA 91355
661-257-8124
airflowresearch.com

BMP
420 Hamilton Rd., Unit A
Deland, FL 32724
386-279-7131
billmitchellproducts.com

Dart Machinery
353 Oliver St.
Troy, MI 48084
248-362-1188
dartheads.com

Edelbrock
2700 California St.
Torrance, CA 90503
310-781-2222
edelbrock.com

Mast Motorsports
330 NW Stallings Dr.
Nacogdoches, TX 75964
866-551-4916
mastmotorsports.com

RHS
3416 Democrat Rd.
Memphis, TN 38118
877-776-4323
www.racingheadservice.com

Trick Flow Specialties
PO Box 909
Akron, OH 44309
330-630-1555
trickflow.com

Rockers

Competition Cams
3406 Democrat Rd.
Memphis, TN 38118
901-795-2400
compcams.com

Crane Cams
1830 Holsonback Dr.
Daytona Beach, FL 32117
866-388-5120
cranecams.com

Crower Cams & Equipment
6180 Business Center Ct.
San Diego, CA 92154
619-661-6477
crower.com

Edelbrock
2700 California St.
Torrance, CA 90503
310-781-2222
edelbrock.com

Erson Camshafts
7301 Global Dr.
Louisville, KY 40258
800-641-7920
pbm-erson.com

Ferrea Racing Components
2600 NW 55th Ct., Ste. 234
Ft. Lauderdale, FL 33309
888-733-2505
ferrea.com

Harland Sharp
19769 Progress Dr.
Strongsville, OH 44149
440-238-3260
harlandsharp.com

Isky Racing Cams
16020 S. Broadway
Gardena, CA 90248
iskycams.com

Jesel
1985 Cedar Bridge Ave., Suite 2
Lakewood, NJ 08701
732-901-1800
jesel.com

Lunati
8649 Hacks Cross Rd.
Olive Branch, MS 38654
662-892-1500
lunatipower.com

Manley Performance Products
1960 Swarthmore Ave.
Lakewood, NJ 08701
800-526-1362
manleyperformance.com

Manton Rockerarms
601 Crane St., Bldg. C
Lake Elsinore, CA 92530
951-245-6565
mantonrockerarms.com

Probe Industries
2555 W. 237th St.
Torrance, CA 90505
310-784-2977
probeindustries.com

Scorpion Racing Products
5817 NW 44th Ave.
Ocala, FL 34482
352-512-0800
scorpionracingproducts.com

T&D Machine Products
4859 Convair Dr.
Carson City, NV 89706
775-884-2292
tdmach.com

Trick Flow Specialties
PO Box 909
Akron, OH 44309
330-630-1555
trickflow.com

Lifters

BAM Products
(Solid Rollers)
Ormand Beach, FL 32174
386-631-6401
hughesengines.com

Competition Cams
3406 Democrat Rd.
Memphis, TN 38118
901-795-2400
compcams.com

Crane Cams
1830 Holsonback Dr.
Daytona Beach, FL 32117
866-388-5120
cranecams.com

Hylift-Johnson
(Hydraulic Rollers)
1185 E. Keating Ave.
Muskegon, MI 49442-6018
800-441-1400
hylift-johnson.com

Jesel
1985 Cedar Bridge Ave., Ste. 2
Lakewood, NJ 08701
732-901-1800
jesel.com

Lunati
8649 Hacks Cross Rd.
Olive Branch, MS 38654
662-892-1500
lunatipower.com

Morel Lifters
John Callies
johncalliesinc.com

Intake Manifolds

Edelbrock
2700 California St.
Torrance, CA 90503
310-781-2222
www.edelbrock.com

FAST
(Fuel Air Spark Technology)
3400 Democrat Rd.
Memphis, TN 38118
901-260-3278
fuelairspark.com

Holley Performance Products
1801 Russellville Rd.
Bowling Green, KY 42102-3542
270-781-9741
holley.com

Katech
24324 Sorrentino Ct.
Clinton Twp., MI 48035
586-791-4120
katechengines.com

Lingenfelter Performance Engineering
1557 Winchester Rd.
Decatur, IN 46733
260-724-2552
lingenfelter.com

Mast Motorsports
330 NW Stallings Dr.
Nacogdoches, TX 75964
866-551-4916
mastmotorsports.com

Water Pumps

Edelbrock
2700 California St.
Torrance, CA 90503
310-781-2222
edelbrock.com

Meziere Enterprises
220 S. Hale Ave.
Escondido, CA 92029-1719
800-208-1755
meziere.com

Moroso Performance Products
80 Carter Dr.
Guilford, CT 06437
203-453-6571
moroso.com

Oil Pumps (Wet Sump)

Federal-Mogul
26555 Northwestern Hwy.
Southfield, MI 48033
248-354-7700
federal-mogul.com

Melling Select Performance
PO Box 1188
Jackson, MI 49204
517-787-8172
melling.com

Pioneer Automotive Industries
5184 Pioneer Rd.
Meridian, MS 39301
800-821-2303
pioneerautoind.com

Oil Pumps (Dry Sump)

AVIAID Oil Systems
10041 Canoga Ave.
Chatsworth, CA 91311-3004
818-998-8991
aviaid.com

Barnes Systems
3162 Kashiwa St.
Torrance, CA 90505-4011
310-534-3844
barnessystems.com

Canton Racing Products
232 Branford Rd.
North Branford, CT 06471
203-481-9460
cantonracingproducts.com

CV Products
42 High Tech Blvd.
Thomasville, NC 27360-5560
800-448-1223
cvproducts.com

Jones Racing Products
72 Annawanda Rd.
Ottsville, PA 18942
610-847-2028
jonesracingproducts.com

Katech
24324 Sorrentino Ct.
Clinton Twp., MI 48035
586-791-4120
katechengines.com

Peterson Fluid Systems
9801 Havana St.
Henderson, CO 80640
800-926-7867
www.petersonfluidsys.com

Timing Sets

Cloyes Gear & Products
6101 Phoenix Ave., Ste. 2
Ft. Smith, AR 72903
479-484-5555
cloyes.com

Competition Cams
3406 Democrat Rd.
Memphis, TN 38118
901-795-2400
compcams.com

Edelbrock
2700 California St.
Torrance, CA 90503
310-781-2222
edelbrock.com

Federal-Mogul
26555 Northwestern Hwy.
Southfield, MI 48033
248-354-7700
www.federal-mogul.com

Lunati
8649 Hacks Cross Rd.
Olive Branch, MS 38654
662-892-1500
lunatipower.com

Milodon
2250 Agate Ct.
Simi Valley, CA 93065
805-577-5950
milodon.com

Ignition

AEM Performance Electronics
2205 126th St., Unit A
Hawthorne, CA 90250
310-484-2322
aemelectronics.com

Edelbrock
2700 California St.
Torrance, CA 90503
310-781-2222
edelbrock.com

FAST
(Fuel Air Spark Technology)
3400 Democrat Rd.
Memphis, TN 38118
901-260-3278
fuelairspark.com

Holley Performance Products
1801 Russellville Rd.
Bowling Green, KY 42102
270-781-9741
holley.com

Katech
24324 Sorrentino Ct.
Clinton Twp., MI 48035
586-791-4120
katechengines.com

Moroso Performance Products
80 Carter Dr.
Guilford, CT 06437
203-453-6571
moroso.com

MSD Performance
1490 Henry Brennan Dr.
El Paso, TX 79936-6805
915-857-5200
msdperformance.com

Valve Covers/Oil Pans

Canton Racing Products
232 Branford Rd.
North Branford, CT 06471
203-481-9460
cantonracingproducts.com

Holley Performance Products
1801 Russellville Rd.
Bowling Green, KY 42102
270-781-9741
holley.com

Katech
24324 Sorrentino Ct.
Clinton Twp., MI 48035
586-791-4120
katechengines.com

Milodon
2250 Agate Ct.
Simi Valley, CA 93065
805-577-5950
milodon.com

Moroso Performance Products
80 Carter Dr.
Guilford, CT 06437
203-453-6571
moroso.com

Stef's Performance Products
693 Cross St.
Lakewood, NJ 08701
732-367-8700
stefsperformance.net

Summit Racing Equipment
1200 Southeast Ave.
Tallmadge, OH 44278
330-630-0270
summitracing.com

EFI Components

Edelbrock
2700 California St.
Torrance, CA 90503
310-781-2222
edelbrock.com

FAST
(Fuel Air Spark Technology)
3400 Democrat Rd.
Memphis, TN 38118
901-260-3278
fuelairspark.com

Holley Performance Products
1801 Russellville Rd.
Bowling Green, KY 42102
270-781-9741
holley.com

MSD Performance
1490 Henry Brennan Dr.
El Paso, TX 79936-6805
915-857-5200
msdperformance.com

Exhaust Headers

Doug Thorley Headers
803 East Parkridge Ave.
Corona, CA 92879
951-739-5900
dougthorleyheaders.com

Hedman Headers
12438 Putnam St.
Whittier, CA 90602
562-921-0404
hedman.com

Hooker Headers
Holley Performance Products
1801 Russellville Rd.
Bowling Green, KY 42102
270-781-9741
holley.com

Sanderson Headers
517 Railroad Ave.
South San Francisco, CA 94080
800-669-2430 or 650-583-6617
sandersonheaders.com

Schoenfeld Headers
605 S. 40th St.
Van Buren, AR 72956
479-474-7529
schoenfeldheaders.com

Stahl Headers
1515 Mt. Rose Ave.
York, PA 17403-2994
855-846-6800
stahlheaders.com

Stainless Works
9899 E. Washington St.
Chagrin Falls, OH 44023
800-878-3635
stainlessworks.net

Ultimate Headers
682 West Bagley Rd., Unit 12
Berea, OH 44017
440-234-9600
ultimateheaders.com

Engine Coating Sources/Services

Cerakote Ceramic Coatings
7050 6th St.
White City, OR 97503
866-774-7628
cerakotehightemp.com

Dart Machinery
353 Oliver St.
Troy, MI 48084
248-362-1188
dartheads.com

Diamond Pistons
23003 Diamond Dr.
Clinton Twp., MI 48035
877-552-2112
diamondracing.net

Federal-Mogul/Sealed Power
26555 Northwestern Hwy.
Southfield, MI 48033
800-325-8885
federal-mogul.com

JE Pistons
8 Mason St.
Irvine, CA 92618
714-895-9594
jepistons.com

Mahle Clevite
1350 Eisenhower Pl.
Ann Arbor, MI 48108
800-338-8786
mahleclevite.com

Polydyn
(Polymer Dynamics)
11211 Neeshaw Dr.
Houston, TX 77065
888-765-9396
polydyn.com

Swain Tech Coatings
PO Box 33
Scottsville, NY 14546
585-889-2786
swaintech.com

Wiseco Pistons
7201 Industrial Park Blvd.
Mentor, OH 44060-5396
800-321-1364
wiseco.com